# Application of Composite Materials

# Application of Composite Materials

Edited by:

Mulmudi Hemant Kumar, Ph.D.

ARCLER PRESS

www.arclerpress.com

**Application of Composite Materials**

*Mulmudi Hemant Kumar, Ph.D.*

**Arcler Press**
**2010 Winston Park Drive,**
**2nd Floor**
**Oakville, ON L6H 5R7**
**Canada**
**www.arclerpress.com**
Tel: 001-289-291-7705
001-905-616-2116
Fax: 001-289-291-7601
Email: orders@arclereducation.com



ISBN: 978-1-77361-094-8 (Hardcover)

# ABOUT THE EDITOR

**MULMUDI HEMANT KUMAR, PH.D.**

Hemant obtained his PhD from Nanyang Technological University, Singapore in 2014. His interests span from material synthesis via solution processed methods to it application in energy harvesting devices such as photovoltaics and water splitting. He is currently working as a research fellow at the Australian national university on silicon/perovskite solar cells.

# TABLE OF CONTENTS

# KIST OF FIGURES

# LIST OF TABLES

# PREFACE

The prospect of achieving any specific set of properties has stimulated the wide-ranging applications of composite materials in numerous industries. Composites, Nanocomposites and Bio-composites are being recognized as superior class of materials for aerospace, sports, transportation, mining, civil, electronics and many other industrial sectors. Fibrous materials have been broadly used in composites in various forms (tows, fabrics, nanofiber webs, sort fibers, mats, etc.) for strengthening purpose or for adding several functionalities. Vast flexibility of structure designing, material selection, and attainable features led to their widespread use in many industrial applications.

This book discusses various categories of composite materials and their applications in modern composite industry. A preliminary discussion on the fundamental characteristics of composite materials, and their manufacturing techniques is illustrated in Chapter 1, i.e. Introduction to Composite Materials. Chapter 2 focusses on applications of composite materials in modern automotive industry. Domestic applications of composite materials are presented in Chapter 3, which focusses on household furniture products. Fiber reinforced composites are extensively employed in construction and civil sectors; detailed discussion regarding building structures and their repairs is presented in Chapter 4. Composite materials have enormous potential to revolutionize the energy sector. Use of advanced composites is discussed in Chapter 5, i.e. Uses of Composite Materials in Energy Sector. The subsequent five chapters, i.e. "Aerospace Applications of Composite Materials", "Military and Defense Applications of Composite Materials", Applications of Composite Materials in Mining Industry", Chapter 9: Uses of Composite Materials in Electronic and Bio sensing Applications" and "Applications of Composite Materials in Sports", cover innumerable categories of composite materials for diverse applications.

This book contains a comprehensive collection of data on all brands of fibers, matrices and composite structures employed to develop a wide range of materials from polymers, metals, ceramics, plastics and natural fibers. Therefore, it is believed that this book can serve as a greatly valuable reference for a varied range of readers, including scientists, engineers, policy makers and professors.

This book is also useful for teachers and students from different backgrounds including textile engineering, materials science, fiber science and technology, mechanical engineering, nanotechnology, chemical engineering, medical sciences and environmental science.

In the end, I would like to extend my sincere gratitude to the editors of this book, and the members of scientific society for their kind support and help throughout the compilation of the book.

Edited

# CHAPTER 1

# INTRODUCTION TO COMPOSITE MATERIALS

## BACKGROUND

In recent years, high-stiffness, high-strength and lightweight composite materials have gained recognition as a significant cutting edge technology with a potential to generate efficient power, re-invent the high efficiency transportation sector, offer new carriages for storage and transportation of reduced carbon fuels, and enhance renewable energy production (Metals & Materials Society, 2012). It is imperative to harness the true potential of composite materials by utilizing the advanced engineering techniques to manufacture the cost-effective yet efficient products at commercial scale. Technology Assessment is an effective tool to modify the manufacturing processes by incorporating the modern advancements in technology.

A composite material is a blend of two or more distinct materials that retain their separate-structure resulting in a unique material which has improved properties than the original constituents (Campbell, 2010). Fiber reinforced polymers (FRPs) are composites prepared by joining a polymer resin with high strength reinforcing fibers. These lightweight composites are used in numerous applications where energy saving and reduced carbon emissions are desired. Major examples of energy conservation include light weight vehicles with lower energy demand, lightweight and high-efficiency wind turbines and utilization of compressed gas tanks for fuel gases.

Normally, a composite material comprises of matrix and fibers. The reinforcement or fiber material gives strength and distributes the applied load in the composite. The matrix material binds and upholds the arrangement or spacing of the fibers and shields the reinforcement from abrasion and environmental damages. The mingling of a matrix with high strength fibers results in the production of a composite material with high strength to weight ratio. The properties of composite materials are often comparable or even superior to the metals. Fibers and resin can be mixed in a significant number of ways and treated through a chain of forming and joining steps. The particular assembling procedure is reliant on the matrix material, the size and shape of the reinforcement, and the basic properties essential for the end use. Based on the nature of matrix material, the composite materials are classified into three categories which are listed below (Adrian & Gheorghe, 2010):

- Organic/Polymer matrix composites
- Metallic matrix composites
- Ceramics matrix composites

The composite materials have following advantages over metallic materials:

- Lightweight
- Low radar visibility
- Easy moldable to complex forms
- Low electrical conductivity and thermal expansion,
- Part consolidation due to lower overall system costs
- Excellent fatigue resistance
- High specific stiffness and strength

As compared to metals, the composite materials also have some disadvantages which are listed below:

- Difficulty manufacturing
- Cost of materials

- Solvent or moisture attack Fasteners
- Long development time
- Temperature limits
- Low ductility
- Hidden damages and damage susceptibility

There are following three kinds of composites according to widespread occurrence:

## Natural Composites

Natural composites are found in both plants and animals. Wood is also a composite – it consists of long cellulose strands (a polymer) held together by a substantially weaker material called lignin. Cellulose is also present in cotton, however without the lignin to tie it together, it is considerably weaker. The two frail substances –cellulose and lignin – combine to form a substantially stronger composite. The bones in our bodies are also composite materials. A bone is composed of a strong yet brittle substance called hydroxyapatite (contains calcium phosphate) and a delicate material called collagen (protein). Collagen is also present in fingernails and hair. Collagen alone is too weak to offer any strength to the bones; however, it can consolidate with hydroxyapatite to provide bones the properties that are expected to reinforce the body.

## Early composites

Human beings have been creating composites since ancient times. One primary example is the manufacturing of mud blocks. Mud can be molded and dried out into a block/brick shape to provide a building material. It has high compressive strength, yet it breaks effortlessly if we attempt to twist it (i.e. weak tensile strength). Straw appears to be extremely strong when we try to stretch it; however, we can fold it up with ease. By combining mud and straw together, it is conceivable to create bricks that are impervious to both crushing and tearing and make fantastic building squares (Makar et al., 2005).

Another old-timer composite is concrete. Concrete is a blend of aggregate (gravel or small stones), sand and cement. It has an excellent compressive strength (it opposes squashing). Recently, it has been observed that adding metal wires or rods to the concrete can enhance its tensile and bending strength. Concrete containing wires or rods are called metal reinforced concrete.

### Modern examples

The primary present day composite material is fiberglass. It is commonly utilized today for watercraft frames, sports gear, building boards and numerous auto bodies. The reinforcement is fiberglass while the matrix is a polymer; the reinforcement has been converted into fine filaments and often woven into a kind of fabric. Discretely, the glass is strong but also very brittle and it will break if twisted sharply. The polymer resin holds the glass filaments together and shields them from impairment by sharing out the stresses being applied to them. Some high-quality composites are presently manufactured by utilizing carbon fibers rather than glass fibers. Carbon fibers are stronger and lighter than glass fibers but more costly to manufacture. They are utilized as components of airplane structures and expensive sports equipment, e.g. golf clubs. Carbon Nanotubes (CNTs) have been used effectively to make advanced composites. CNTS based composites are stronger and lighter than the composites manufactured from conventional carbon fibers. They do, nonetheless, offer prospects for building lighter aircraft and cars (which will utilize less fuel than the currently used heavier vehicles).

The world's biggest airliner, Airbus-A380, contains modern composite materials in its structures. More than 20 % of the Airbus-A380 structure is made of modern composite materials; mostly plastic strengthened with carbon fibers. The Airbus design utilized fiberglass-reinforced aluminum on a large scale for the first time in the history of air crafts. The novel design is 25% stronger and 20% lighter than the conventional aluminum airframe.

## FIBER REINFORCED PLASTICS (FRPS)

FRPs for wind turbine blade, compressed gas storage, and automotive applications are considered as primary models for clean energy applications. There are different other applications of FRPs including industrial equipment such as pipelines and heat exchangers, geothermal power generation, structural building materials, flywheels for stability of electricity grid, hydrokinetic energy production, support assemblies for solar systems, transportation containers and different frameworks which can benefit from cost effective high strength, corrosion resistant, and lightweight composites to influence national energy objectives. Some of these applications benefit mainly from CFRP-based composites, which provide a high stiffness-to-weight ratio and excellent strength-to-weight ratio as compared to numerous structural materials (Tannous & Saadatmanesh, 1998). These lightweight composite materials can provide significant energy conservation and saving during the utilization stage that can't be accomplished with materials that don't possess high stiffness and strength properties. Comparison of properties for different categories of materials is shown in Fig.1.

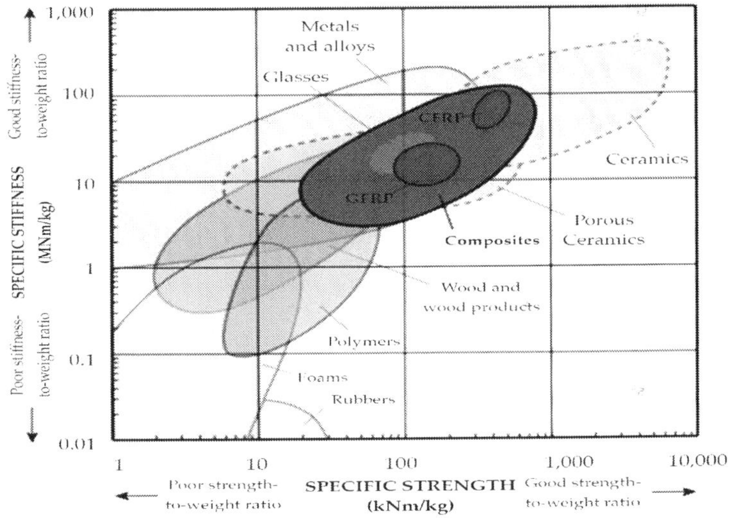


**Figure1:** Specific strength and specific stiffness for different materials, the figure also highlights Glass Fiber Reinforced Plastic (GFRP) and Carbon Fiber Reinforced Plastic (CFRP) Composites.

## SCIENCE OF COMPOSITE MATERIALS

The reinforcing component provides the stiffness and strength in a composite material. In most of the cases, the reinforcement material is harder, stiffer and stronger than the matrix phase. The reinforcement is typically a particulate or fiber. Dimensions of particulate composites are nearly equal in all material directions. They might be platelets, spherical or any other symmetrical or asymmetrical geometry. Particulate composites have a tendency to be significantly less stiff and weaker than continuous fiber reinforced composites, yet they are less costly. Particulate strengthened composites generally contain less particulate content (40 to 50 % volume) due to brittleness and processing difficulties. The length of a fiber is significantly greater than its corresponding diameter. The length-to-diameter ratio of a fiber is called the *aspect ratio* (*l*/*d*) and can vary prominently (Halpin, 1969). Discontinuous fibers have small aspect ratios while continuous fibers possess comparatively higher aspect ratios.

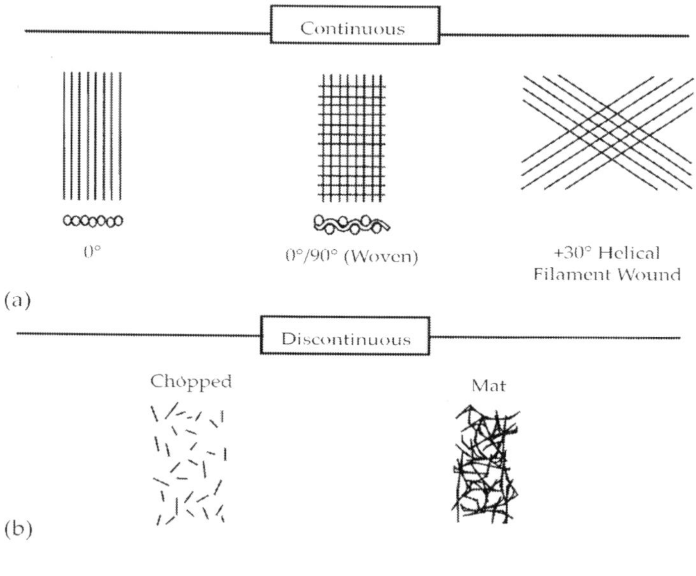


**Figure 2:** Typical reinforcement types

Continuous fiber reinforced composites typically have a regular fiber orientation, while discontinuous fibers composites generally have an irregular fiber orientation in the matrix. Continuous fibers include unidirectional fibers, helically winded fibers and woven cloth (Fig. 2a), while discontinuous reinforcements are characterized as a random mat and chopped fibers (Fig. 2b).

Continuous fiber reinforced composites are generally converted into composite laminates by stacking continuous fiber sheets in various orientations to achieve the desired stiffness and strength properties with high fiber volumes, i.e. 60 to 70 percent. Fibers deliver high-strength composites on account of their smaller diameters; they contain far fewer defects (typically surface imperfections) as compared to the components manufactured by bulk materials. The diameter of fibers is inversely proportional to the strength of a composite, i.e. smaller the diameter, the greater its strength, however, smaller fiber diameters result in cost increment of the composite materials (Migneault et al., 2008). Additionally, high strength fibers with a smaller diameter are highly flexible and more responsive to manufacturing operations such as forming or weaving. Typical examples of the fibers include glass, carbon, and aramid which may exist as both continuous and discontinuous fibers. The continuous phase in a composite is the matrix, which is either a polymer, ceramic or metal.

Polymeric matrix has low stiffness and strength, metallic matrix possesses mild stiffness and strength but high ductility, and a ceramic matrix is brittle in nature but has high stiffness and strength. The matrix phase performs numerous important functions such as keeping the fibers in a proper alignment and shielding them from environment and abrasion. In metal and polymer matrix composites which make a solid bond between the matrix and the fiber, the load is transmitted from the matrix to the fibers through shear loading at the fiber-matrix interface. In ceramic matrix composites, the goal is to enhance the toughness of the composite rather than stiffness or strength. Therefore, a composite with a low interfacial (fiber-matrix interface) bond strength is desirable. Quantity and the type of the reinforcement define the final characteristics of a composite. Fig 3 demonstrates that the higher modulus and strength values are achieved with continuous-fiber composite materials. There is a practical limit (about 70 percent volume) for fiber content in a composite. Fiber content beyond that limit results in loss of strength and mechanical properties due to the inability of the matrix to support and hold the fibers properly.

The strength of the composite with discontinuous fibers can match that of continuous-fiber composite if the aspect ratios of discontinuous fiber composites are sufficiently high and they are regularly aligned. However, it is difficult to keep up a substantial arrangement with discontinuous fibers. Discontinuous-fiber composites are usually irregular in alignment, which significantly decreases their modulus and strength. On the contrary, discontinuous fiber composites are typically less expensive than the continuous-fiber composites. Consequently, continuous-fiber composites are utilized where higher stiffness and strength arc required (quality driven), and discontinuous fiber composites are utilized where cost is the principal driver while stiffness and strength are less significant. Nature of both the fiber and the matrix influence processing parameters for composite manufacturing. There are two categories of polymer matrices, i.e. thermoplastics and thermosets (Blackketter et al., 1993). A thermoset is initially a low-viscosity resin (matrix) that reacts with environment and cures during the processing, resulting in a rigid solid.

A thermoplastic is a highly viscous resin which is prepared by heating it over its liquefying temperature. Since a thermoset resin sets up and eventually cures during the processing phase, it can't be recycled by reheating. There are processing methods for both categories of resins which are more amiable to continuous fibers while there are other processes which are more suitable for discontinuous fibers. Ceramic and metal matrix composites are usually more expensive than polymer matrix composites because of high processing temperatures and pressures. On the other hand, both metal and ceramic matrix composites possess higher thermal stability, which makes them an ideal candidate for high-temperature applications. Variation in strength for the different class of fibers is shown in Fig. 3.

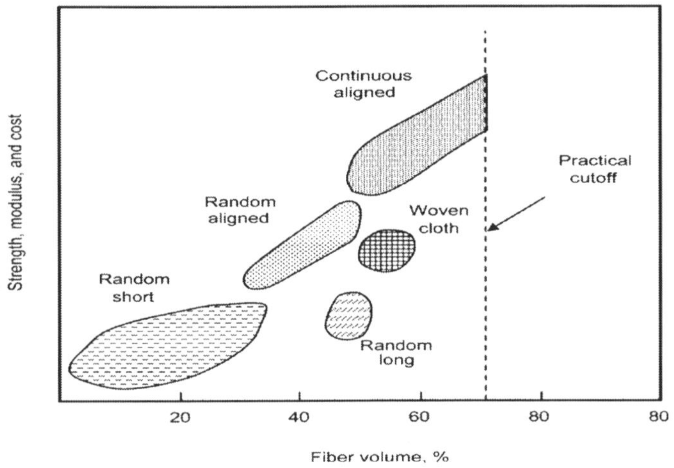


**Figure 3:** Influence of reinforcement type and quantity on composite performance

# ORTHOTROPIC, ISOTROPIC AND ANISOTROPIC MATERIALS

All the materials are categorized as either isotropic or anisotropic material (Jones, 1975). Isotropic materials possess the similar properties in all material directions, and perpendicular loads generate normal strains only. However, anisotropic materials usually have dissimilar properties in all material directions. There are no symmetrical planes in anisotropic materials, and normal loading conditions produce both normal and shear strains. Materials are categorized as isotropic if their properties are not direction dependent within the bulk material. For instance, if the material undergoes loading along 0°, 45°, and 90° from a specific reference point, the modulus of elasticity remains same in all directions, i.e. $E0° = E45° = E90°$. Different loading conditions for an isotropic material as explained in Fig. 4.

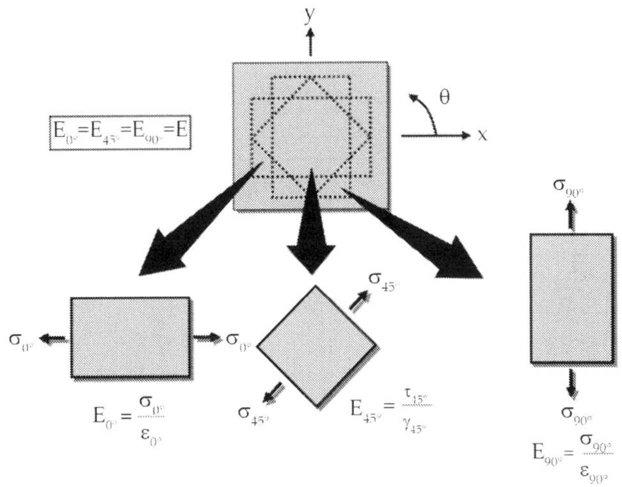


**Figure 4:** Isotropic material under stress conditions

## Element of isotropic material under stress

The properties of anisotropic materials vary significantly with change in material direction. The modulus of the anisotropic material also varies with the change in direction, i.e. $E0° \neq E45° \neq E90°$. Variation of other mechanical properties such as ultimate strength, thermal expansion, and poison's ratio are also direction dependent in anisotropic materials. Bulk materials (metals, polymers, and ceramics) are typically dealt as isotropic materials, and composite materials are classified as anisotropic materials. However, bulk materials, for example, metals and polymers can also become anisotropic, i.e. if these materials are treated (cold working, localized heating) to harvest certain properties in specific directions.

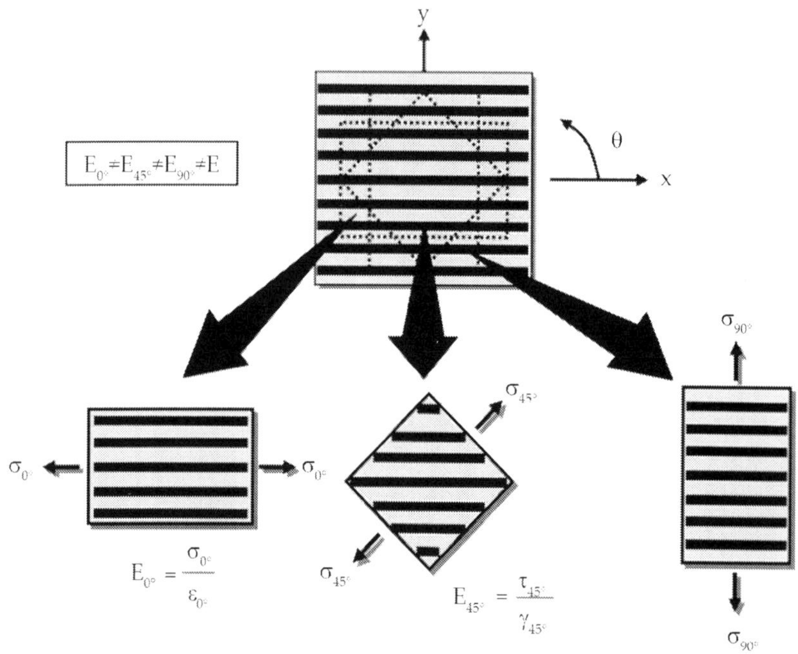


**Figure 5:** Anisotropic material under stress conditions

## Elements of composite materials under stress

Composites materials lie in the sub-category of anisotropic materials which are branded as orthotropic materials. Properties of orthotropic materials are different in all perpendicular directions. Their three axes of symmetry are mutually perpendicular, and only normal strains are produced by application of load parallel to these symmetrical axes. On the contrary, both shear and normal strains are produced if the applied load is not parallel. Therefore, orthotropic properties of the material are regulated by material orientation.

# RULE OF MIXTURES AND CURING OF COMPOSITES

The longitudinal modulus of a continuous fiber laminate undergoing unidirectional loading, parallel to the fibers (11- 0° direction), can be calculated from the individual properties of constituents by utilizing Rule of Mixtures (Hui & Shia 1998).

## Rule of Mixtures

$$E_{11} = E_f V_f + E_m V_m \qquad \text{(eq 1.1)}$$

In equation 1.1, $E_f$ is the modulus of fiber, $V_f$ is volume percentage of fibers, $E_m$ is the modulus of the matrix, and $V_m$ is the volume percentage of the matrix. The longitudinal ultimate tensile strength ($\sigma_{11}$) can also be calculated by applying the rule of mixtures:

$$\sigma_{11} = \sigma_f V_f + \sigma_m V_m \qquad \text{(eq 1.2)}$$

In equation 1.2, $\sigma_m$ and $\sigma_f$ are the ultimate strengths of matrix and fiber respectively. As the strength of fibers is extremely high as opposed to the strength of matrix at all fiber volume percentages, therefore, the strength values of matrix can be ignored:

$$E_{11} \approx E_f V_f \qquad \text{(eq 1.3)}$$

$$\sigma_{11} \approx \sigma_f V_f \qquad \text{(eq 1.4)}$$

## Curing Processes

Structures composed of fiber-reinforced plastics (FRPs) necessitate the polymer matrix (resin) to attain and sustain solid-state properties in service. Thermoset resins polymerize irreversibly through cross-linking reaction, while thermoplastics are easily re-melted above their transition temperature. Thermoplastics and thermosets not only differ in physical properties but also undergo different fabrication processes for composite manufacturing. Traditionally, advanced composite materials have long been based on thermosetting polymer systems, and almost 80% of the composites are based on a thermoset polymer matrix, which requires a particular curing time to produce the desired properties.

Due to stringent certification and specifications methods, aerospace composite materials are based on epoxy resin systems in which the curing reactions follow a specific temperature profile to ensure a proper resin flow, consolidation, de-gassing, and ultimately homogeneous degree of polymerization to attain final properties (U.S. Department of Energy, 2011). The curing processes are usually slow (a few minutes to hours) and energy exhaustive because the huge thermal mass of autoclave and the tooling are also subjected to the same thermal sequence. Autoclave methods have also been utilized by the composites industries beyond aerospace applications. The modern developments and improvements in optimized cure cycles, selective polymerization techniques, and further innovations in the autoclave technique are prospective steps to diminish the energy consuming composite manufacturing processes.

Currently, electro-technologies using radiation heat transfer principles are being utilized for selective heating or curing of composites materials. However, electro-technology requires composite systems which are responsive to applied radiation frequencies. Examples of responsive systems are explained below:

Dielectric heating techniques based on radio frequency (RF) or microwave (MW) where the electromagnetic (EM) radiation combines primarily with the matrix; for instance, the RF curing of epoxy-based composites are dependent on the dielectric response of the resin system (Feher et al., 2003).

- Infrared (IR) heating is a cost effective and efficient technique of heating, pre-heating, melting and curing. Medium and long-wave IR heating has various potential applications; which have been successfully implemented by composite manufacturing industries such as partial curing of composites and pre-heating of fiber pre-forms during intermediary processing steps.
- Induction heating techniques are utilized to heat conductive matrix materials such as metallic matrices. Induction heating processes are broadly used in metal foundries for heat treatment and melting purposes (Cresko, & Roberts, 2002).

## COMPOSITES VERSUS METALLIC MATERIALS

As discussed earlier, the physical and mechanical characteristics of metals and composites are substantially different. Table 1 depicts the comparison of properties for composites against metals.

**Table 1:** The comparison of properties for composites as compared to metals (Campbell, 2010)

| Condition | Comparative behavior relative to metals |
|---|---|
| Load-strain relationship | More linear strain to failure |
| Notch sensitivity | |
| Static | Greater sensitivity |
| Fatigue | Less sensitivity |
| Transverse properties | Weaker |
| Mechanical property variability | Higher |
| Fatigue strength | Higher |
| Sensitivity to hydrothermal environment | Greater |
| Sensitivity to corrosion | Much less |
| Damage growth mechanism | In-plane delamination instead of through thickness cracks |

Comparison of the through-the-thickness mechanical strength of a composite laminate and an aluminum alloy is displayed in Fig.6.

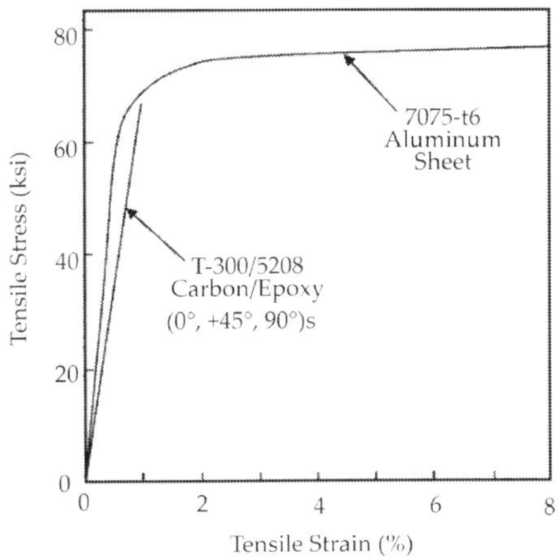


**Figure 6:** Comparison of through the thickness strength for epoxy composite and aluminum alloy

Composite materials are exceptionally anisotropic, in-plane stiffness and strength of composites are typically high and vary in different directions, depending on the alignment of the fibers. There are some composite properties which do not show any improvement because of reinforcement, i.e. through-thickness tensile strength. The load is applied on the matrix rather than fibers in this thickness direction.

Ductility of metals is typically high which makes them compress or elongate significantly, without fracture, under certain loading conditions. Ductile yielding has following two benefits:

- Ductile yielding assists in local stress relief by distributing the stress to adjacent structure or material. Consequently, ductile metals possess a huge capacity to incorporate stress relief mechanism during static loading conditions.
- Ductile materials have a great capability to absorb energy (area under the stress-strain curve). Therefore, ductile metals undergo plastic deformation under impact loading before fracture.

Composite materials are brittle as compared to metals. Brittle nature of the composites is characterized by its poor capability to withstand stress concentrations. Brittle composites also lack the ability to withstand impact loads. The response of notched composite materials to cyclic (fatigue) loading is also considerably different (high) from that of metallic materials. The capacity of composites to tolerate fatigue loading is greater than that of metals. However, the static strength of notched (damaged) composites is lower than that of metals due to the presence of defects.

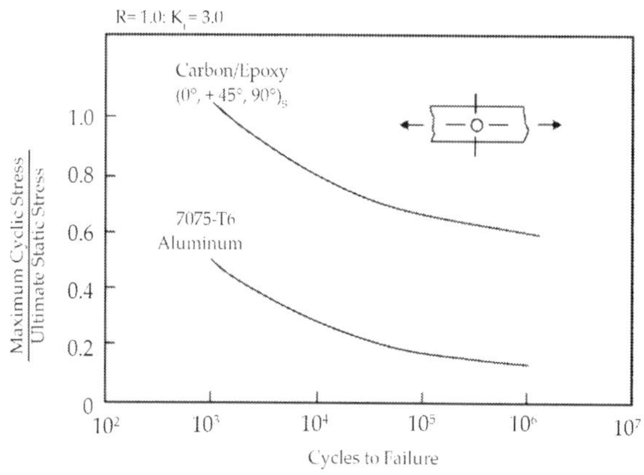


**Figure 7:** Comparison of Fatigue properties for epoxy composite and aluminum alloy

Fig.7 shows that the fatigue strength of a notched composite material (carbon/epoxy composite) is considerably higher than that of space grade aluminum alloy (7075-T6). Structural materials usually require high static strength as compared to high fatigue resistance; therefore, the strength of the composites is improved by adjusting the fiber content and its orientation. However, fatigue is generally a critical design parameter.

## ADVANTAGES AND DRAWBACKS OF COMPOSITE MATERIALS

There are many advantages of composite materials such as lighter weight, improved fatigue life, high corrosion resistance, high stiffness and strength, and lower assembly expenditures due to lesser part and fastener components (Campbell, 2010). Specific modulus (modulus/density) and specific strength (strength ratio density) of highly strong fibers are generally greater than the mechanical properties of comparable aerospace alloys (Fig. 8). Higher mechanical composite materials are also lighter in weight, which result in improved performance, longer range, greater payloads, and fuel saving. Fig. 8 depicts the comparison of structural efficiency for carbon/epoxy composite, high strength aluminum alloy (7075-T6) and titanium alloy (Ti-6Al-4V).

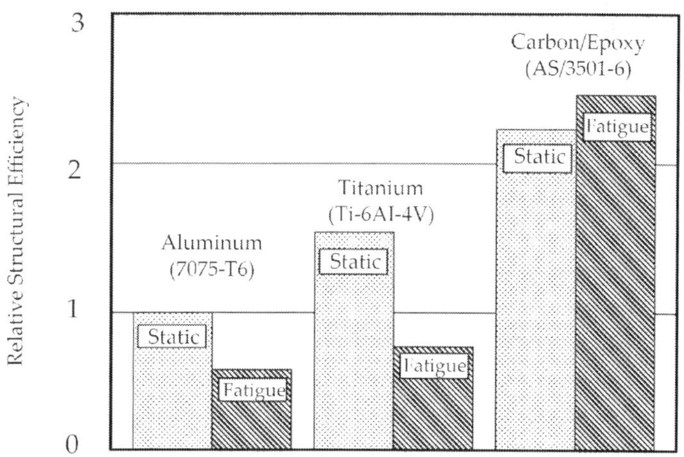


**Figure 8:** Structural efficiency of different materials

# FABRICATION PROCESS FOR COMPOSITE MATERIALS

Composite fabrication normally accounts for manufacturing of following two structures (Strong, 2008):

### *Laminates*

Some composites are fabricated as Laminates which contain bonded layers.

### *Sandwiches*

Composite materials are also manufactured as honeycomb sandwich structures which consist of a low-density core sandwiched between faces of polymeric or metallic sheets.

Composites manufacturing can be carried out by various processes, some important fabrication processes are discussed in the appending text. Flowchart of the composite manufacturing processes is shown in Fig. 9.

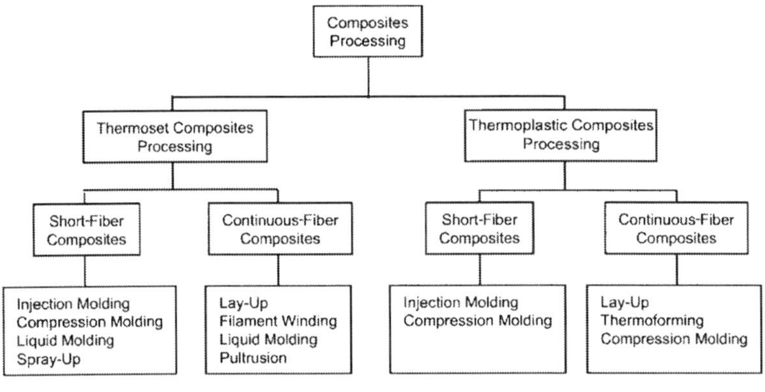


**Figure 9:** Major polymer matrix composite fabrication processes

## Injection Molding

Injection Molding is one of the most well-known and broadly utilized fabrication processes for high-volume development of thermoplastic matrix components reinforced with fibers (Thiriez & Gutowski, 2006). Almost 20% of all products fabricated these days injection molding because of its versatility, adaptability and lower cost. Solid blocks of resin (matrix) containing reinforcing fibers are fed into the heated barrel (containing rotary screw) through a hopper. The rotary screw produces heat shearing against the barrel, ultimately liquefying (melting) the resin. The rotating screw also behaves like a piston which forces the blend of molten resin and fibers into a metal mold, which aids the mixture to cool down and solidify.

After solidification process, the composite component is ejected by opening the mold cavity. The major benefits of injection molding include the short process durations (generally, a few seconds) and comfort of automating the manufacturing process. The combination of process automation and short process times results in higher production volumes. The principle drawback of injection molding includes the high cost of the molds and capital equipment. Additionally, the properties of composites also vary due to the absence of a mechanism for distribution and orientation of fibers. Furthermore, because of the limitations in the melt viscosity of thermoplastic resins, injection molding accounts for the production of particulate fiber reinforced composite materials which are utilized in the manufacturing of automotive components, such as dashboard components, seat backs, closures and control valves.

Long process durations for injection molding of parts are a major downside to the utilization of fiber reinforced plastics in high volume marketplaces, including typical automotive applications. Long process cycles are regulated by curing rate, resin flow rate and the time required to circumvent the formation of bubbles in the polymeric resin which converts into cavities after curing and result in structural flaws. Considering the competition in the market, the cycle times of composites manufactured from injection molding are typically less than two minutes which are considerably faster than the conventional pre-preg processes (usually one-hour cycles). Existing composites applications normally utilize glass fibers and a thermoplastic resin (instead of thermoset resins) for manufacturing of injection molded components with shorter cure times.

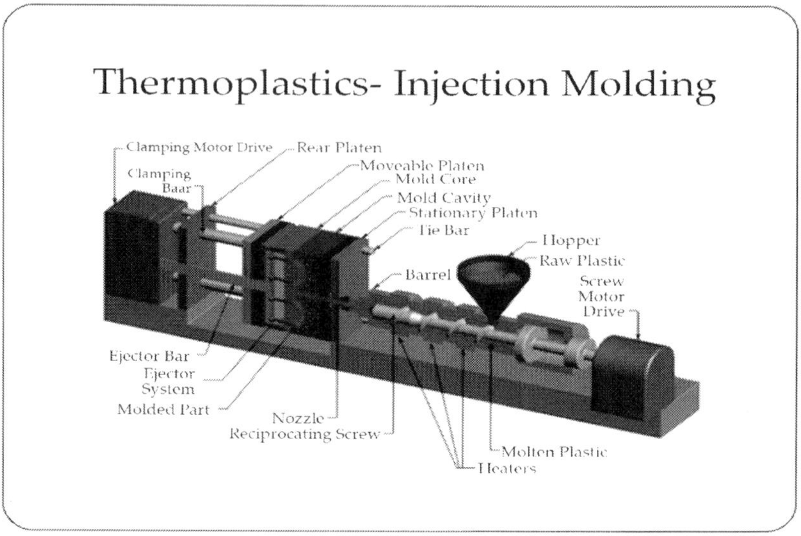


**Figure 10:** Schematic of Injection Molding process

Substantial advancements are in progress to regulate the chemistry of thermoplastics and tailor the viscosity of resins to manufacture continuous fiber reinforced composites. This high volume fabrication process will remain the process of choice for fabrication of automotive components. Recently, Toho Tenax has developed a technology based on carbon fiber reinforced plastics which can produce high volumes of automotive components in less than a minute.

## Resin Transfer Molding (RTM)

Resin Transfer Molding involves the packing of dry fiber reinforcement of fiber preform into a mold tool which has the desired geometry of the composite component. Both mold tools are clamped over one another and the resin in fed into the mold cavity (Potter, 2012). A vacuum process known as Vacuum Assisted Resin Injection might be utilized to facilitate the drawing of resin through the cavity. The key drawbacks of this process include the inability of the tool to manufacture large-sized components and high costs of mold tools, which can withstand high pressures. Moreover, un-impregnated zones can exist resulting in extremely expensive scrap parts. This composites fabrication method offers the most astounding capability of all fabrication processes in the manufacturing of intricate, large-scale automotive parts. The current BMW i3 utilizes RTM method in combination with the robotic alignment of fiber preforms to fabricate the frame of the car. The process is also a potential fabrication method for chassis/suspension and hood applications in auto vehicles.

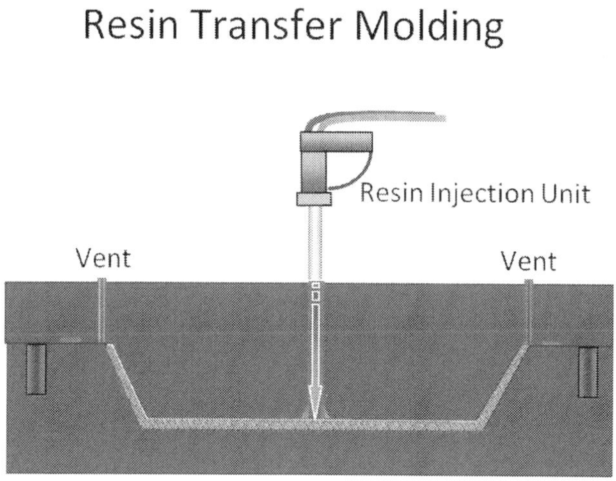


**Figure 11:** Schematic of Resin Transfer Molding

The development of rapid curing epoxy and polyurethane resins are the key manufactured high-performance carbon fiber composites formed by utilizing RTM, infusion molding or compression molding processes. High-pressure RTM in conjunction with thermoforming is a promising latest development to improve the process cycle of the RTM procedure. Presently, the process time of RTM is approximately 20 minute with the utilization of high-pressure infusion of resin to reduce the injection time to seconds rather than minutes and takes into account the utilization of highly reactive thermoset resins. All the major worldwide providers of thermoset resins have formulated lab-scale resins having less than two-minute process durations. Currently, Scaling up the RTM procedure for high-pressure infusion and quick curing resin systems is the prospective challenge which is being addressed.

## Vacuum-Assisted Resin Infusion

There are many minor variations in Resin Transfer Molding (RTM) which involve the replacement of the upper mold tool by a vacuum bag. These modified procedures include RIFT, SCRIMP, and VARTM. A permeable layer, for instance, a knitted non-structural fabric, is usually used to facilitate the dispersal of the resin in part (Goren & Atas, 2008).These advanced processes have substituted Resin Transfer Molding (RTM) for many applications due to their simplicity, low capital cost, and the capability to fabricate huge material structures such as rail carriages and bridge sections. The major drawbacks of these modified processes include the poor surface finishing of the product on the bagging side, longer times consumed in material preparation, limitation of the processes to flat structures, lack of automation and poor dimensional tolerance. Wind turbine blades are being extensively manufactured by vacuum assisted resin transfer molding which contains low curing temperature pre-preg with 90 % glass fibers. Besides high labor content, the process cycles for VARTM are typically very high spanning 35-40 hours for a 45m wind turbine blade. Automated fiber placement (AFP) and inspection are essential to reduce the overall labor content. The use of thermoplastic resins will substantially reduce the cost, blade weight, cure cycle durations, and facilitate the processing of composites after their expiry date. Currently, a unique automated method for manufacturing of wind turbine blades, from glass fabric layup solution, is being developed. Due to high cost, carbon fiber use has been limited to spar cap applications today. Using pultruded reinforcements of carbon fiber sheets in wind blade spars has proved to assist in developing lighter and larger rotors which will improve energy capture. This novel process is suitable for manufacturing larger wind blades (~100 m long) wind blade without the requirement of autoclaves. Similar to RTM, extensive research is being carried out to improve the VARI (vacuum assisted resin infusion) process by developing fast curing resins to condense the cycle times.

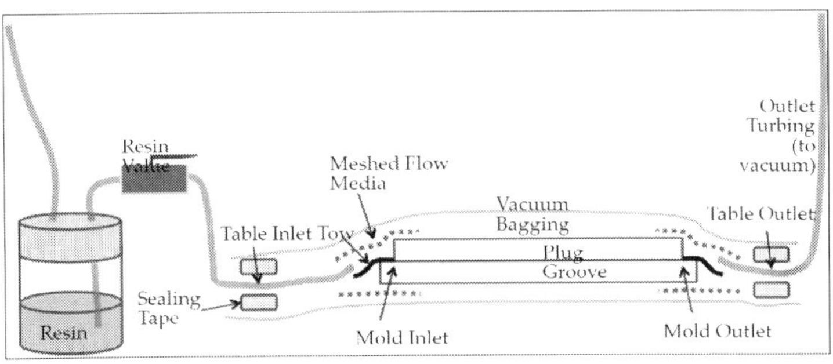


**Figure 12:** Schematic of Vacuum Assisted Resin Infusion process

## Compression Molding

The working principle of compression molding is typically simple that has been used for decades. The source material (charge) is poured into the cavity of a mold. The material charge is a blend of fibers and resin. Typical pressure up to 2000 psi is applied to force the charge to adopt the shape of the mold cavity. Molding carried out at lower pressures is known as cold press molding. The benefits of compression molding are its simplicity, high repeatability, fast cycle times, high-volume production and precise tolerances. The major drawbacks include the huge initial capital costs in presses molds, minor defects as a consequence of residual stresses, warpage, delamination, and flow orientation of the fibers (Isayev, 2012). Compression molding process is presently being utilized in non-structural automotive applications, for instance, closures, interiors, and miscellaneous components. The primary source materials for compression molding process are short fiber reinforced bulk molding compounds (BMCs) and sheet molding compounds (SMCs).

Modern compression molding processes are being developed which incorporate long carbon fibers for reinforcement in SMCs and BMCs to achieve improved stiffness and strength-to-weight ratios. Research is also being conducted to toughen the composite materials for enhanced crack resistance.

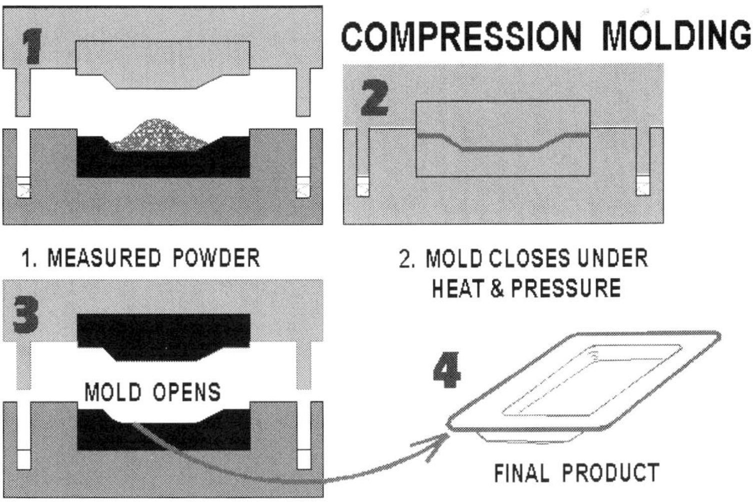


**Figure 13:** Schematic of Compression Molding process

Currently, most of the composite manufacturing industries are also manufacturing their indigenous resins and SMCs to meet the requirements of specific applications, i.e. high UV resistance, improved moisture and impact resistance, enhanced surface finish, etc. A sub-category compression molding process called Matched Die Molding has a strong potential to manufacture continuous fiber reinforced components for structural automotive applications such as chassis, car body and suspension. In Matched Die Molding, the fiber ply sheet (blank) is compressed into the final shape in die-mold assembly and cured rapidly to produce final components. The blank design is engineered accurately to accommodate the alterations in fiber orientation during the processing. The cure cycle time normally depends on the matrix material; i.e. thermoplastic parts are consolidated in a few seconds while thermosetting parts are cured in in minutes (typically 17 minutes). As mentioned earlier, extensive research is underway to formulate thermoset resins with lesser cure times (approximately 2 minutes). Compression molding process is competitive to Resin Transfer Molding (RTM) if the dies remain reusable after multiple attempts without any significant geometric distortions.

## Hand Lay Up

In hand lay-up method, the resin is manually impregnated by hand into fabrics and fiber weaves. Brushes or roller are normally used for hand layup method before the composite is set to cure under normal atmospheric conditions. The main disadvantage of hand layup method is the lack of consistency in the product; the product quality is greatly dependent on the laminator skills. Low viscosity resins are typically used in hand layup process. This practice usually compromises the thermal and mechanical properties of the composite material and causes health hazards for the laminator.

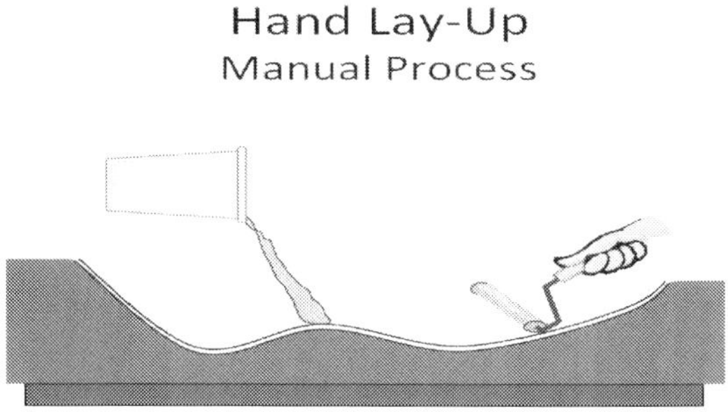


**Figure 14:** Schematic of Hand-Layup method

## Spray Up

In a spray up process, catalyzed resin and chopped fibers are showered into a mold where curing takes place under ordinary environmental conditions. Although the spray up process is less costly; it tends to cause serious damages. Laminates have a tendency to be enormously resin-rich and in turn become extremely heavy. Only low viscous resins and short fibers are easily sprayed in this process which seriously limits the improvement in mechanical properties. The utilization of high styrene resins can possibly be dangerous.

**Figure 15:** Schematic of Spray-Up process

It is big challenge to manage the volatile organic compounds (VOCs) and harmful air pollutants which are produced during the process. Due to the high costs involved in control mechanisms of these harmful products, most of the composites companies have shifted to infusion-based and closed mold processes, which are good at managing the pollutants. Composite parts manufactured from the advanced process have an excellent surface finish and accurate dimensional control which cannot be attained with hand layup or spray-up processes. Therefore, these conventional methods are only being used for repair of the damaged parts of concrete, steel, and plastics.

## Filament Winding

Filament winding process is most suitable for manufacturing hollow, oval or circular parts, such as tanks and pipes. Fiber tows are wound on a rotary mandrel after passing through the resin bath. The main drawbacks of the filament winding process include low viscosity of the resin and inability of the fibers to be laid in an axial direction. This is a major composite manufacturing method for axisymmetric composite materials, such as pipeline sections and gas storage tanks. Filament winding process offers lower process time and cost efficiency for structural axisymmetric components such as drive shafts, struts, and axles. For high volume manufacturing of storage tanks by using carbon fibers and epoxy resin, the cost of carbon fibers normally contributes to 60% of the total cost incurred in the filament winding process. Reduction in process cost and modification of the process cycle time to manufacture 500,000 parts per annum can be attained through the use of innovative braided preforms, low energy cost by shorter cure times, lower raw material cost, reduced scrap, manufacturing automation, and use of shielding coatings and mechanically durable materials.

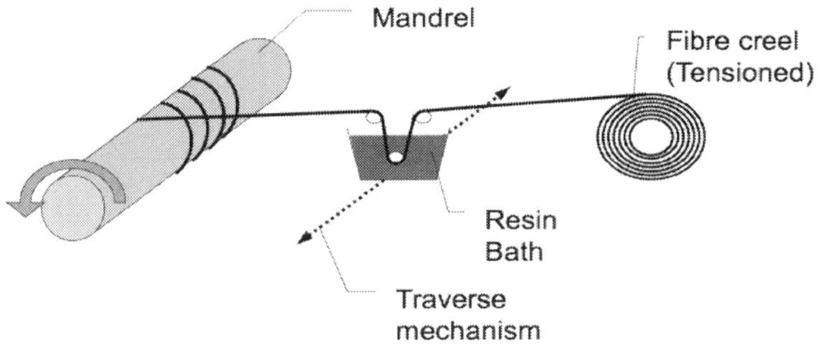


**Figure 16:** Schematic of Filament Winding process

## Pultrusion

In pultrusion process, the fibers are pulled from a fiber roll and then dipped in a resin bath. The fibers pass through a heated die which cures the resin. This process is suitable for manufacturing of parts with nearly constant cross-sections. Moreover, the initial capital cost of the heated dies is typically high. Pultrusion produces even finished components that normally do not involve any post processing (Gorthala et al., 1994). A broad variety of continuous, solid, hollow and consistent profiles are pultruded easily, and the process is flexible enough to be easily tailored for manufacturing of specific products, such as the cross-section spars for wind blade applications.

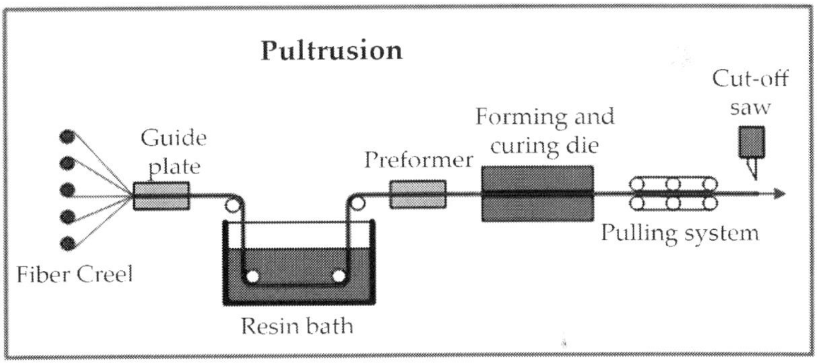


**Figure 17:** Schematic of Pultrusion process

## AUTOMATED FIBER PLACEMENT (AFP)

Automated tape laying (ATL) and automated tow placement (ATP) are sub categories of this process. The major difference between the two processes is the nature of starting material and the fiber laydown rates. The fiber tow placement method automatically places various separate pre-preg tows onto a rotary mandrel at extraordinary speeds. This process uses a numerical controlled robotic placement head to clamp, distribute, dissect and restart up to 32 fiber tows simultaneously. The shape of the final composite blank is highly dependent on the length of the shortest fiber tow, which can be laid down by the machine. The fiber or tow placement head can be fixed to a 5-axis gantry support, which is retrofitted to a filament winding system.

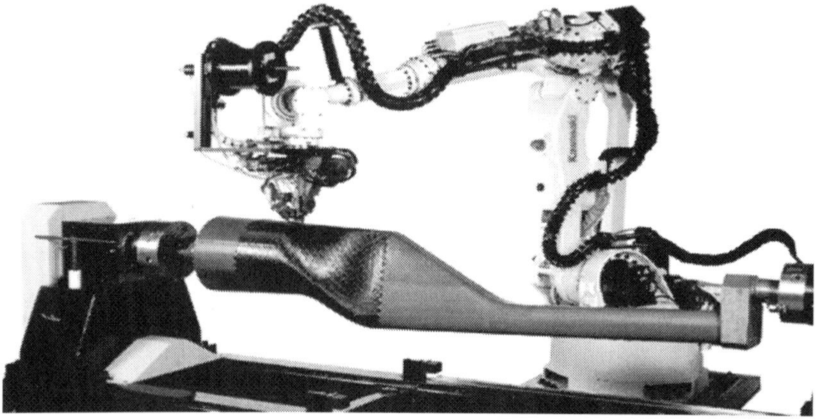

**Figure 18:** AFP setup for composite manufacturing

Automatic tow placement machines are typically available with twin mandrel stations to enhance the productivity. Benefits of fiber placement are high processing speed, consolidation of parts and enhanced part-to-part homogeneity and lesser labor costs and material scrap. ATP is often employed for the manufacturing of large and intricate thermoset parts (Oldani & Jarvi, 2008). Automated tape laying (ATL) is a fast, automated technique which involves the continuous laying of pre-preg tapes instead of single fiber tows. Components having highly complicated angles or contours are normally manufactured by ATL. Tape layup is a versatile technique, allowing pauses and directional changes in the process. ATL can be utilized for both thermoplastic and thermoset materials. The head of ATL contains a winder, spools of tape, a compaction shoe, winder guides, a tape slitter and a position sensor. The head of ATL can be positioned on the edge of a multi-axis robot which moves around the mandrel onto which the material is being spooled, or the head can be positioned on a gantry hanging above the tool or mandrel. On the contrary, the mandrel can also be rotated to enable the head to access and scan different areas of the mandrel. Fiber or tape is fed to the tool in sequences at any specific length or angle. The sequence typically comprises of one row of the source material. Multiple sequences are usually fed together for any defined area or pattern. The feeding sequences are designed and controlled by a software code which is derived from the design and analysis of part models. Monetary costs for computer-controlled, automated equipment are usually high.

## REFERENCES


1. The Minerals, Metals and Materials Society (2012). Materials: Foundation for the Clean Energy Age.
2. Structural Composite Materials. Campbell, F.C. (2010) ASM International
3. Adrian, P. O. P. P., & Gheorghe, B. M. (2010). Manufacturing Process and Applications of Composite Materials. Volume IX (XIX).
4. Makar, J., Margeson, J., & Luh, J. (2005, January). Carbon nanotube/cement composites-early results and potential applications. CONFERENCE ON CONSTRUCTION MATERIALS.
5. Tannous, F. E., & Saadatmanesh, H. (1998). Environmental effects on the mechanical properties of E-glass FRP rebars. ACI Materials Journal, 95(2), 87-100.
6. Halpin, J. C. (1969). Stiffness and expansion estimates for oriented short fiber composites. Journal of Composite Materials, 3(4), 732-734.
7. Migneault, S., Koubaa, A., Erchiqui, F., Chaala, A., Englund, K., Krause, C., & Wolcott, M. (2008). Effect of fiber length on processing and properties of extruded wood-fiber/HDPE composites. Journal of Applied Polymer Science, 110(2), 1085-1092.
8. Blackketter, D. M., Upadhyaya, D., King, T. R., & King, J. A. (1993). Evaluation of fiber surfaces treatment and sizing on the shear and transverse tensile strengths of carbon fiber-reinforced thermoset and thermoplastic matrix composites. Polymer Composites, 14(5), 430-436.
9. Jones, R. M. (1975). Mechanics of composite materials (Vol. 193). Washington, DC: Scripta Book Company.
10. Hui, C. Y., & Shia, D. (1998). Simple formulae for the effective moduli of unidirectional aligned composites. Polymer Engineering & Science, 38(5), 774-782.
11. U.S. Department of Energy (2011). Industrial Technologies Office Report DOE/EE-0389. Retrieved from http://www1.eere.energy.gov/manufacturing/intensiveprocesses/pdfs/eip_report.pdf
12. Feher, L; Flach, A.; Nuss, V.; Pozzo, P.; Seitz, T. (2003). HEPHAISTOS - A novel 2.45 GHz Microwave System for Aerospace

Composite Fabrication. 9th International Conference on Microwave and R F Heating, Loughborough University, Loughborough.

13. Cresko, J.W; Roberts, P.L. (2002). The method of induction curing conductive carbon fiber composites with radio frequency energy. 3rd World Congress on Microwave and Radio frequency applications, Sydney, 2002.
14. https://inis.iaea.org/search/search.aspx?orig_q=RN:35028342
15. Structural Composite Materials. Campbell, F.C. (2010) ASM International, 10-11.
16. Structural Composite Materials. Campbell, F.C. (2010) ASM International, 14-15.
17. Strong, A. B. (2008). Fundamentals of composites manufacturing: materials, methods, and applications. Society of Manufacturing Engineers.
18. Thiriez, A., & Gutowski, T. (2006, May). An environmental analysis of injection molding. In Electronics and the Environment, 2006. Proceedings of the 2006 IEEE International Symposium on (pp. 195-200). IEEE.
19. Potter, K. (2012). Resin transfer moulding. Springer Science & Business Media.
20. Goren, A., & Atas, C. (2008). Manufacturing of polymer matrix composites using vacuum assisted resin infusion molding. Archives of Materials Science and Engineering, 34(2), 117-120.
21. Isayev, A. (2012). Injection and compression molding fundamentals.
22. Gorthala, R., Roux, J. A., & Vaughan, J. G. (1994). Resin flow, cure and heat transfer analysis for pultrusion process. Journal of Composite Materials, 28(6), 486-506.
23. Oldani, T., & Jarvi, D. (2008). U.S. Patent No. 7,407,556. Washington, DC: U.S. Patent and Trademark

# CHAPTER 2

# USES OF COMPOSITE MATERIALS IN AUTOMOTIVE INDUSTRY

## BACKGROUND

The use of composite materials in automotive industry dates back to 1953, when a beautiful convertible car with white color and red interior was produced and named, 'Chevrolet Corvette'. For the first time in history, structural polymer composites were used in the production of any car. The body of Chevrolet Corvette was made from fiber glass. Although plastics like Bakelite had been in use since 1940 for the production of steering wheels and distributor caps, however, this was the first time that composites were used for the production of any car, and therefore it is considered an important milestone in the history of automotives. From then onwards, fiber toughened polymers have been in continuous use for various automotive applications. Although, certain properties of composites like their reliability, light weightiness, and their ability to get mold in different shapes makes them an ideal substitute for metals, still, due to challenges like precise material characterization, their production, painting and combination with metals – they are not widely used in automotive industry. These issues thus needs to be resolved if composites are to be used in place of metals in the automotive sector. For this purpose, European Commission sponsored a networking initiative named as COMPOSIT project. The main objective of this project was to examine the prospective use of composite materials in automotive industry. Under this project, a series of workshops were organized to discuss ten main points relating to composite design, use, repair, durability and production in the field of transport, rail and aerospace industry.

* Previously, wood which is a natural composite, was used for the manufacturing of vehicles but wood cannot be used to get high volume production.

Top experts from approximately 220 organizations were invited to attend and participate in these workshops. More than 550 people from 17 countries attended these workshops. Attendees included suppliers, designers, researchers, processors, software engineers, consultants, and end users of composite materials.

The workshops held under the COMPOSIT thematic network addressed specific problems related to the consumption of composite materials in transport industry by making comparisons, cooperation and cross reproduction between diverse sectors. Major challenges were discussed which needs to be addressed for using the composites more extensively. Moreover, recommendations were also made to face these challenges in future prospective projects.

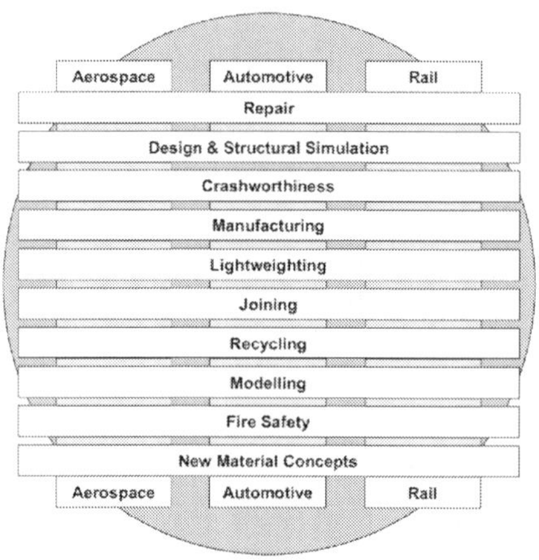


**Figure 19:** Overview of the areas covered by COMPOSIT network

Main recommendations made in the report are stated below:

- For mass production, prevailing composite materials as well as their application processes needs to be modified.
- Fully mechanized manufacturing processes of composite materials needs to be developed.
- Cost of composite materials needs to be reduced.
- Innovative mathematical models needs to be developed for composite materials.
- Special focus needs to be placed for the development of automotive industry related composite materials.

This chapter presents a summary of the findings of COMPOSIT thematic network in the transport industry. It further gives an overview of the current applications of composite materials. Major challenges of using composites in vehicles have been discussed and future prospective applications of composites within the field of automotive industry have been suggested and their feasibility analyzed.

## PRESENT APPLICATIONS OF COMPOSITES IN AUTOMOTIVE INDUSTRY

### Composites in Mass Production Cars

Since 1953, when composites were used for the first time in automotive industry, their applications have been increasing in the transport industry. However, the use of this technology in the mass production sector is still limited due to multiple reasons. Composite materials have lengthy processing times, their raw materials which include fibers, resins, etc. are comparatively costly, which makes it hard to attain good quality finishes of their surfaces (Holbery & Houston, 2006). Further, due to issues like rigorousness and durability of structural composite materials, they have narrow use in high volume car productions. Therefore, majority of vehicle manufacturing companies prefer to use steel.

Presently, thermoplastics are the most widely used polymer materials in transport components. Thermoplastics are often strengthened by mixing them with short glass fibers. High production yield is achievable with the help of these materials by employing process of injection molding and its consequent procedure. For instance, the succession time of thermoplastic dashboard is generally less than 90 seconds. Some other common applications of thermoplastics include their use in making trims of doors, lights, engine manifolds and interior apparatuses. Recently, the use of structurally long fiber reinforced thermoplastics is getting popular. An example of this is the manufacturing of support lid which has been used on the Volvo V70 XC AWD station wagon. Previously used aluminum frame in this component has been replaced by glass reinforced polypropylene. It is anchored directly to the rear axle of wagon and is used for supporting the rear differential. Weight of the part is 2 kg which is almost 27 % less than its aluminum counterpart. An added advantage is the reduced manufacturing cycle time which is now less than 4 minutes.

**Figure 20:** Application of composite support lid in Volvo V70 XC AWD

Bumper beam of BMW M3 is another example where structural thermoplastics have been used. Using this component has enabled manufacturers to reduce the weight of car by 60 %. It also increases the levels of energy absorption (Allison & Cole, 1993).

Another kind of polymers known as thermosetting polymers are not extensively used, and present estimates have shown little opportunities of their use in future. Till now, only two processing methods have been used for these kind of polymers making the process suitably automatic to yield high volume automotive applications.

- Sheet molding compound (SMC)
- Resin transfer molding (RTM).

A high volume vehicle which employed these technologies for the first time is Renault Espace whose 1984 – 1996 version used RTM tailgate and SMC body panels made of polyester. Now, the production level of Renault Espace is around 70,000 vehicles per year, which makes it one of the most substantial application of SMC in automotive industry.

**Figure 21:** Composite panels in Renault Espace

Some other example of SMC use in automotive industry include their use in the manufacturing of vehicles like Fiat Coupe, Mercedes-Benz CL-500, Volvo V70, Alfa Romeo, 2002 Ford Thunderbird and many more. In many of the aforementioned vehicles, it would have been tough to accomplish the desired style by using metals.

## Composites in Sports Cars

As mentioned earlier, the use of composite materials in mass production vehicles is limited. On the other hand, they have been widely used in manufacturing of sport cars. Properties of composite materials like light weightiness, high stiffness, high cost and low production volumes makes them an ideal material to be used in the production of sports cars (Friedrich

& Almajid, 2013). Presently, Porsche Carrera GT and the Mercedes-Benz McLaren SLR, uses carbon fiber reinforced composites in their body structures, whereas, Aston Martin Vanquish V12 uses carbon fiber in their substructure.

**Figure 22:** Composite body of Dodge Viper SRT10

Another example is 2003 model of Dodge Viper SRT-10, which uses glass reinforced reaction injection molded (RIM) wings, a glass reinforced SMC bonnet, and carbon reinforced SMC wing supports, door panels, and windscreen frames. Another popular example is Ferrari Enzo (2003) which uses a carbon reinforced chassis and body. The only metallic modules in this car are some engine supports and crash energy absorbers.

## Prototype Composite Cars

An interesting method which automotive industry uses to explore prospective uses of composites in future is through concept cars and model vehicles. An example is Ultalite concept, manufactured in 1992, which used carbon SMC body. Another example is 2000 models of BMW Z22 which used an RTM single piece carbon fiber reinforced side frame.

**Figure 23:** Composite prototype of BMW

Dodge Viper SRT -10 is yet another example of a composite concept car. This car has employed multiple clear-coated carbon fiber modules which includes a boot lid (using an in-built spoiler to increase back downforce), the bonnet, and a bolt-on hard roof top. Using these composites, total of 70 kg weight has been reduced.

## Composites in Trucks and Trailers

For several years, the composite materials like glass fiber reinforced SMC has been in use for variant applications such as in the manufacturing of body panels, bumpers, rood caps, and step fenders of Lorries, and trucks.

Likewise, SMC bonnet, door panels, bumper and roof cap of the Ford Aeromax 9500, presents a whole of 204 kg of composite materials in this vehicle. Some lorry manufacturers likes to employ thermoplastic resins for their taxis. A PBT resin known as Xenoy has been used in molding the outer panels of Mercedes Benz Actros, Volvo FH and Scania Series 4.

In November 2000, Hendrickson International and Delphi Automotive Systems presented a prototype of a lorry bumper fully made of composite materials. The weight of the bumper was around 10 kg, which was almost 50 kg lighter than its steel models. Large woven tow carbon fiber fabric and glass fibers were employed to make this composite bumper. Within the molding of the bumper, were all the fittings and internal metallic beam assimilated.

**Figure 24:** Trailer container made of composite materials

Composite materials are also used in the manufacturing of trailers. An example is a Swedish refrigerated semi-trailer known as 'The Hycoprod' which employing vacuum infusion uses fiberglass and foam sandwich panels in its manufacturing. Contrary to a traditional design, total 1100 kg of weight is saved by using these materials. A developed prototype has magnificently concluded 250,000 km test track.

Recently, a non-refrigerated composite design has also been manufactured known as the "Composit-trailer". Moreover, composite materials have significantly reduced the weight of trailer containers (Jambor & Beyer, 1997).

## TECHNICAL ISSUES CONCERNING THE USE OF COMPOSITES IN AUTOMOTIVE INDUSTRY

The following sections presents an overview of the issues discussed in series of workshops held under the COMPOSIT project.

### Repair

Main features of automotive industry regarding repair of vehicles components are stated as follows:

- For repairing purposes, vehicles like trucks and cars can be carried to workshops easily.
- Repairing workshops are usually accessible in any area.

- Presently, repairs of composites generally include panels and bumper repairing, some supports and chassis material.
- Components made of composites are easy to disassemble since their weight is much smaller as compared to the metallic components.

As can be analyzed from the points above, repairing of composites in automotive sector normally do not call for any significant issues. Panels and bumpers of vehicles are designed such as to resist any collision during low speed, and effects caused by hitting stones or due to weight of a leaning person. Bumpers usually break if encountered with high energy collisions. Normally, it is cheaper to replace a bumper instead of repairing it which is quite costly. This situation is converse in sport cars. The composite chassis used in the manufacturing of sports cars is generally manufactured in a single country and then distributed throughout the world. Often the repair of the chassis needs an extensive knowledge of used materials and components, therefore, in case of extensive damage, the vehicle is normally required to be returned to the manufacturer for repairing. This issue is resolved much easily in case of limited damage, where parts be replaced in a local workshop. Nevertheless, the repair of non-replaceable parts will become inevitable if composites are to be used extensively in future applications (Thornton & Jeryan, 1988).

## Design and Structural Simulation

During early days, designers did not appreciate the use of composite materials in automotive industry. Conventionally, most designers used isotropic sheets of metals in manufacturing of vehicles. An extensive knowledge of both the composite materials and their associated processes is required to fully exploit the properties of composite materials. Vehicle designers must understand manufacturing processes of composites and their relation with the components being developed. In the early days, composite materials have often been used in a poor manner (Baillie, 2005).

Major reason behind poor processing of composites was a huge gap between the anticipations of composite manufacturers, and vehicle designers which was also a reason of delays in production. Main challenges faced by the automotive industry these days are shortage of characterization of composites, and non-availability of simulation tools. One more problem is the lengthy computational time requisite to model components and structures of composite materials. The lengthy processing time of composite models in most of the commercially designed software usually takes days especially for initial cycle of vehicle development, in which multiple options needs to be analyzed within just few months. For proper evaluation of composite materials at early stages, a reduction factor of 10 in solution times must be observed by the automotive industry. Till now, this problem has not been solved by software developers. Consequently, modern research and design centers are now employing in-house developed methodologies, and keep them confidential. To summarize the issue, it can be said that although modern automotive designers have the knowledge of composite materials yet there is a lack of proper simulation tools needed to simulate the processes during design phases (Mangino et al., 2007).

## Crashworthiness

Initially, target of the car designers was to produce stiff and seemingly non-deformable cars. Gradually, crashworthiness has been developed as a well advanced field over the recent years. Failure of the vehicle's non-deformable chassis to disintegrate the energy resulted in a strong impact received by passengers. With the evolution of crashworthiness strategies, special focus is placed now to manufacture highly deformable vehicles that have expectable energy absorbing zones both at the rear and front end of vehicle. A firm safety shell for passengers is placed in between the front and rear end of vehicle. Moreover, interior of the vehicle has also been made crashworthy by putting air bags, removing sharp edges, making surfaces smooth and using collapsible components. Trucks and trailers haven been redesigned to make them less intrusive in case of impact with any other vehicle. Rules regarding the protection of pedestrians have also been revised to reduce accidents and injuries caused as a result of it. More technical solutions are still under the development phase. Modern solutions include putting external air bags, manufacturing bonnets made of energy absorbing materials, and restyled fronts of vehicles. Crashworthiness of composite materials is much more than metals due to their energy absorbing properties (Carruthers et al. 1998). Experimental procedures to carry out full scale testing for the development of such vehicles is quite costly. Special high standard testing facilities are needed for this purpose. Vehicle under test undergoes inevitable damage and thus cannot be used afterwards. For this purpose, crash simulation tools have been in use since their development (Kurtaran, et al. 2002). For instance, early models used for simulation in case of frontal crashes composed of a comprehensive mesh of front section and the remaining vehicle shown with the help of beam elements and concerted masses.

Advance numerical models used today reproduce complete details of a vehicle, also counting passenger dummies. Detailed analysis like this usually takes days to solve. Commercially developed software used for crash analysis of composite material models usually requires very wide material property data. Although many mathematical material models have been developed, still the lack of reliable data of composite materials causes problems for analysts. To develop an understanding of basic crash behavior of composite materials, researchers are trying to address the issue but only at the micro scale. An example is the lengthy investigation carried out to understand the performance of textile reinforcements. As the requirements generate with time so does the introduction of new materials in markets take place. As far as the future development of composites is concerned, major focus should be placed on the upgradation of simulation tools to make them precise and affordable according to the needs of vehicle designers.

## Manufacturing

Raw materials used for manufacturing of composite materials are costly; therefore composites are not extensively used when high production volume is required (Hashim et al., 1999). The second more important reason of lack of extensive use of composites in the automotive industry is the deficiency of any proper manufacturing process. Manufacturing process is chosen according to the required production rate. Statistics shows that for cars, 80,000 – 500,000 parts per year are manufactured, whereas, trucks and trailers have much less volume of between 5000 and 20,000 parts per year. Some other important aspects to consider include scrap production, cycle time and costs of tools. Composite processes are mostly just single shot operations which is the reason why their processing tools are much cheaper as compared to the tools used for metal sheet forming, which requires 5 – 6 separate tools for each module line. For low production volumes, savings due to lesser tool costs seems convincingly influential, whereas, for high production volumes, the dominant factor is the cost of parts making composites less desirable in such cases. Currently, for high production volumes, the only used composite manufacturing processes are short fiber reinforced thermoplastic injection molding and bulk moulding compound (BMC) processes.

Major benefits of injection molding are:

- Very short succession time
- Little resultant scrap is produced.

Presently, there are not many processes for medium volume composite production. The two most popular methods used are:

- Compression molding by using sheet molding compound (SMC)
- Glass mat thermoplastic (GMT)

Both of the above mentioned processes are fully automatic with succession times of few minutes. These methods are used for cars and trucks. Problems like high density, paint-ability, and surface finishing which were faced originally with these materials have now been resolved. The problem of scrap production and requisite post machining still persists in case of SMC as well as GMT. The bumper of any truck if produced using SMC, will require the milling of holes for its assembly over the vehicle. As a result scrap will be produced which must be disposed of carefully. For medium volume production, a promising composite material is fiber reinforced thermoplastic which has adequate stiffness as well as shorter succession period.

Another method used for medium volume composite production is resin transfer molding (RTM). Use of RTM is getting popular in structural applications due to its higher potential for automation, higher tolerances and attainable mechanical properties (Potter, 2012). An additional advantage of using RTM parts is high quality surface finishes. Main disadvantages of using this process include high amount of produced scrap, high tooling costs and lengthy succession times.

## Light weighting

Light weighting is one of the main reason why composite materials are getting popular in the automotive industry. The advantages of light weighting differs for different kinds of vehicles. For example, light weighting in trucks have enabled us to increase our payload without increasing the overall mass (Shankar & Vijayarangan, 2006).

In sports vehicles, by reducing weight of cars, an increase in acceleration and top speed is achieved. Less amount of fuel is consumed by light weighted vehicles, which translates into reduced emission of $CO_2$ and other gases. However, most of the car manufacturers do not consider development of low emission hybrid electric/ gasoline engines enough to meet regulatory targets as stated by government. Use of zero emission engines is very limited in routine applications. Hence, it is expected that by employing lightweight materials and their respective designs, environmental obligations of automotive industry can be fulfilled.

Some other factors which need to be taken care of in light weight design are:

- An optimal negotiation among weight saving and extra cost
- Standard passive safety measures
- High level surface finishes
- Comfort of passenger should be taken into account
- For white component bodies, verified manufacturing methodologies

Certain internal industry aspects like payload and public laws play an important role in driving the applications of light weight design in transport industry. Concept cars have been used to demonstrate lightweight materials and their associated manufacturing processes. Presently, the main obstacles that need to be resolved are high costs of raw materials and manufacturing productivity (Jones, 1998).

## Joining

In designing and manufacturing of a vehicle, joining is a critical phase. Each joint disrupts the structural geometry which gives birth to number of problems like causing discontinuities in materials, problems in transferring load, and local peak stresses. It also increases noise and vibration produced by the vehicle.

### *Hybrid material design*

Current trend is to unite materials and their processes in a synergistic manner which is a reason why these issues are aggravating. A more effective technology for using composite materials in automotive industry is "Hybrid material design". In this design procedure, materials are selected based on their individual properties and then united on functional basis that will exploit their properties in a best possible manner.

There are two main advantages of using this method:

- Weight of vehicle is reduced.
- Structural performance can be increased.

However, performance of Hybrid structures containing unlike materials rely on their joining technologies. These structures must be again separated into their ingredient parts to recover their materials [14]. Three main classes of joints are:

- Mechanical fastening
- Adhesive bonding
- Welding / fusion bonding.

Hybrid designs (for example using adhesive bonding and fastening simultaneously) are also in use (Ashby & Brechet, 2003). The most appropriate and in practiced method of joining is mechanical joining through bolts and rivets. Some advantages of using mechanical joints are listed below:

- No need of final finishing
- Easy reassembling and disassembling of parts for purposes of examination or maintenance.
- Mechanical joining usually takes less time
- Environmental factors such as temperature do not severely affect these joints.

However, this method has also some disadvantages, some of which are mentioned below:

- Increase in weight
- Stress concentration
- Requirement of overlapping parts
- Condition of high tolerances

- Likelihood of galvanic corrosion.

Unluckily, the practices followed for mechanical joining of metallic structures do not work same for composite materials. Main reasons behind this are:

- Cuts and holes drilled causes stress concentration, and stress redistribution of composite materials is limited due to deficiency of plasticity in them.
- When localized drilling is carried out, composite materials begin to delaminate.
- Thermal expansion properties of composites differ from those of metallic fasteners.
- There exists a likelihood of galvanic corrosion.
- Extra weight is added due to mechanical joints which work against the light weighting properties of composite materials.

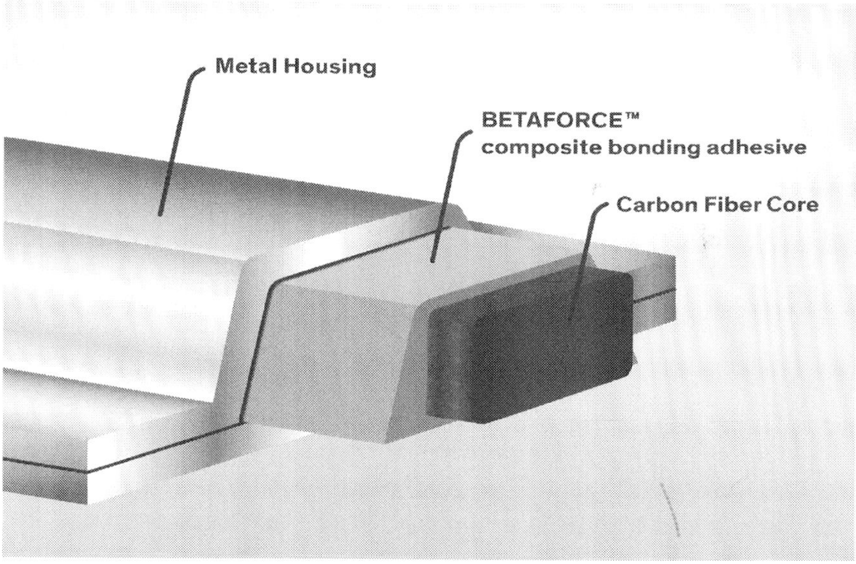


**Figure 25:** Adhesive bonding of composite materials

## *Adhesive bonding*

Popularity of adhesive bonding is increasing from some years now, owing to the availability of high performing polymer based adhesive materials. This process has many positive attributes including:

The mechanical properties of adhesive materials can be altered as per design requirements

- Adhesive joints have higher exhaustion and corrosion properties which increases the life span of products using them.
- Adhesives provide brilliant sealing
- Smooth surfaces between two fused materials provide an added styling advantage.
- Discontinuities caused by drilling of holes can be avoided by distributing load transfer pathway over the complete joined area. In this way peak stresses are reduced.
- By taking advantage of gap filling properties of adhesives, there is a fair chance of increasing limits of tolerances.
- Flexibility of complete system can be enhanced by tuning stiffness of adhesive materials
- Adhesives can be applied using fully automatic systems
- Vehicle's noise and vibration features can be improved by adhesive bonding.

Some of the disadvantages of this technology are:

- Surface needs to be prepared before processing
- Thermal distortion caused as a result of high temperature curing
- Autoclave curing

### *Welding*

Welding technology is well established technology used in automotive sector for amalgamation of metallic parts. However, it has not been widely used with composites.

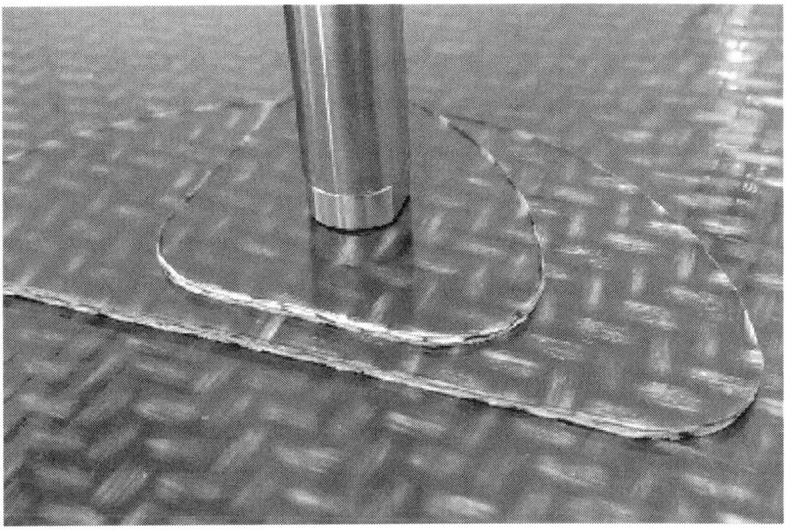

**Figure 25:** Composite welding and joining process

- Main benefits of welding are:
- High mechanical performance
- Robustness of joints
- Easy on-line examination
- Short processing times
- Trifling surface preparation

Three main techniques which are used for welding of thermoplastics are induction, ultrasonic, and resistance welding.

Main drawbacks of welding include:

- Complex assembly
- Limitation to thermoplastics
- Inductive welding needs conductive fillers
- Resistance welding requires metal meshes
- All three techniques require quantity of carbon fiber content to be low.

Future prospective joining techniques of composites are not mature yet. Although there are some exciting and convenient technologies, but their long succession times restrict them to be used only in low production volumes.

Welding techniques are becoming more sophisticated with time by improving their automation and processes. In case of composites however, they are not as much practicable as they are with mechanical joints.

In automotive sector, the most progressing technology is adhesive bonding. Future focus of adhesive development is to increase their functionality, improving their adhesion to low energy surfaces, expansion of hybrid adhesives, and better ecological credentials. Regardless of the future joining techniques of composites, a major requirement is to develop and introduce new rules and mathematical methods for the design, simulation and testing of joints.

## Recycling

For many years, recycling methods have been in use in automotive industry. In early times, metallic material of cars, buses, Lorries, trucks and etc., was recycled by melting their metallic structure. In this way virgin raw material was obtained which was also economically viable. Metals can be easily recycled since they deform easily every time when appropriate heat is applied to them. Composites on the other hand, behave differently.

To enable easy recycling, automotive industry is reducing use of different polymers. At the end of a vehicle's life, plastic components used in it are separated. Nevertheless, with every passing year, number of cars which need to be recycled are increasing. Presently, 75 % of a car's total weight is due to metallic parts, and almost 25 % is due to non-metallic components including rubber, plastics and glass.

Today, the bigger problem in recycling of composite materials is not the recycling technology itself, but the cost and use of the material after being recycled. Total cost after recycling of composite like fillers, and reinforcements is much higher as compared to their virgin counterparts. Technical performance of recycled composite materials cannot be guaranteed as the standard of their quality is doubt-full unlike virgin materials. Resultantly, very few automotive parts are presently contrived from the recycled composites.

Problems in recycling of composites arise due to the fiber and resin material in them which remains discreet. Now a days, self-reinforced materials have been started to develop to get rid of this problem. A number of natural fiber reinforcements like coconut, basalt, feathers of birds, and abaca are used industrially for applications in cosmetics. These materials are safe to use from recycling prospective as they can be destroyed easily without producing any residues as by product. Still, lot of development is required in the applications of long fiber reinforcements.

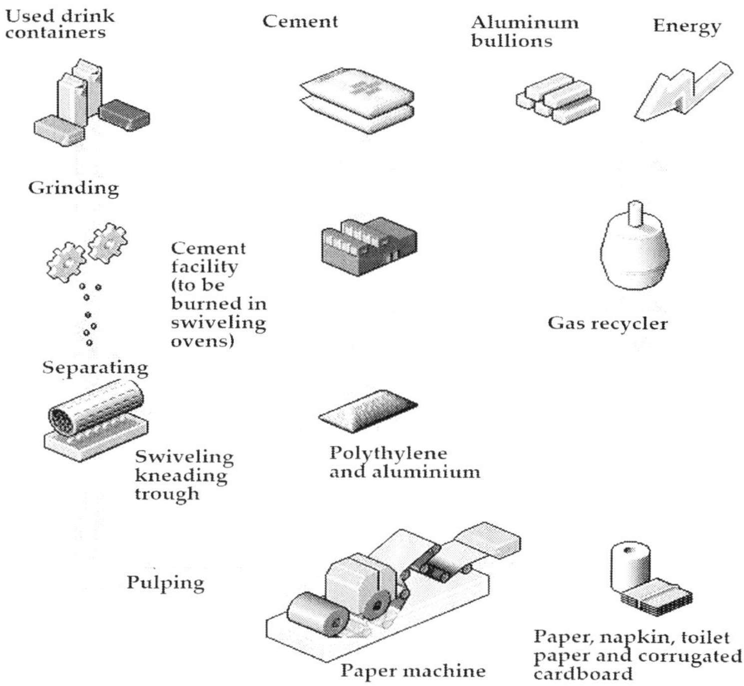


**Figure 27:** Recycling cycle of typical composite products

Care should be taken as to not create an ever increasing pile of raw materials after recycling. For instance, according to theory, short fiber reinforced thermoplastics can be recycled without any difficulty by melting them (Yang et al., 2012). Laboratory experiments have proved that short fiber reinforced thermoplastics can be grinded many times without any significant loss in their structural performances. However, reality on roads is in contrast to lab results. Life span of a car is normally 17 years during which it is constantly exposed to harsh environmental factors like sunlight, pollution, acid rains and etc. As a result, resinous materials degrade with passing time which affects their properties after recycling. Thus only a fraction of recycled material is added to raw virgin material for developing new materials. A new design approach should be adopted for recycling procedures. Materials and components used should be carefully chosen and designed so as to ease their separation afterwards. They should be categorized according to the parameters like re-usage, recycling and energy reclamation. Further, processes for recycling should be developed. Car designers have still not solved the problem of recycling multi-material hybrid modules. Two methods are in practice these days:

- Shredding of the components
- De-assembling of components

Research should be taken to investigate recycling methods for structures composed of hybrid materials. A better approach might be to combine the expansion of modern recycling methods and approaches used by other industrial sectors that are facing the analogous constraints.

## Modelling

Two important phases of automotive sector are modelling and mathematical simulation. They are required to lessen the marketing time of new products and costs related to investigational testing. Simulation is carried out in two main areas:

- Design of vehicle
- Manufacturing processes

Automotive industry is carrying out structural analysis of vehicle design for years now. Although models have matured and are more accurate now, but until recently only metals and fraction of polymers have been involved in this. With the increase in use of composites in sports cars and other vehicles, more rigorous models of composites are needed to develop (Simacek et al., 2008).

During modeling of composite materials, one main issue is keeping balance between intricacy of model and suitable computational solution time. In the initial design phase, it is absolute to complete analysis of as many as one million element models per single day. However, this target is impossible to achieve in current situation. The situation is even worse in case of multi objective optimizations or while interrogating disparities in properties of materials due to variations. Both methods need minimum of total 60 – 100 simulations. One day is normally required to run each analysis which means it will take more than 3 months to complete the whole process.

While dealing with the issues related to modeling of composites, it is preferable to begin with micro models of fabrics, and fiber reinforced composites, to gain a deeper perceptive of the physics of the situation. Next comes the stage of developing an intermediate model. Final goal is to develop a macro model presenting precise characterizations with the help of comparatively small amount of material data. In practiced composite manufacturing processes like RTM, SMC, GMT have been maintained by numerical formulas and therefore can be simulated. For high volume production, repetitive optimization must be ensured to reduce the recycling time, amount of scrap produced, and number of discarded parts.

For material characterization, new materials are needed to create which is not an easy task as material characterization is costly and needs a lot of time. There are few commercial models available in market who model reinforcing fabric drawn arranged according to its final shape. Environment of workshop is a limitation as well. There is always a chance of error in case of fabric positioning performed by operator instead of doing it mechanically. Accurate modeling is therefore hard to achieve. Verification of performance of any product can be carried out by integrating the design and analyses procedure. This is of utmost importance for designers who want to maintain structural characteristics of material while reducing the weight to minimum. Prospective future development in composite field is expected in areas like damage modeling, reducing solution times, improvement in failure theories and etc.

## Fire Safety

As compared to rails or airplanes, cars have fewer chances of getting caught into fire emergencies mainly because of their comparatively low travel speed, feasibility of stopping by whenever needed, and possibility of easy evacuation in case of emergencies. Due to these reasons, automotive regulations for fire safety are less severe than in other transport means. If use of combustible materials are to be increased then automotive fire regulations must also be paid attention (Chapple & Anandjiwala, 2010).

Evacuation can be easier for a healthy person, but if someone is severely wounded, or old, or a child unable to move by himself, then it may take too long for them to escape from a burning car. These factors cannot be ignored. Presently, regulation in practice is ECE34, which legalizes tests related seepage of fuel from tanks. It outlines features of a fuel tank; mentioning maximum allowed seepage in case of impacts on front, back or any side; considering it as a stand-alone component as well as part of an obstructed vehicle.

## Innovative Material Conceptions

It is not possible to predict materials which will be used in coming years for the automotive sector. Presently, we are in an intermediate phase, with manufactures pursuing innovative products, or to make use of prevailing technologies used by other industrial fields. Regardless of future prospective materials, a new method to evaluate materials is getting popular. This approach is known as life cycle analysis (LCA). In this approach, all phases of product's life cycle are taken into account, from production of raw materials to product's end of life phases including its recycling and energy recovery. Composites produced from natural resources are interesting from life cycle analysis viewpoint. However, it is not always necessary that natural will mean environmental friendly. For instance, plant oil is used to produce polymers which are biodegradable and can be genetically engineered in order to improve their characteristics. Nevertheless, several negative aspects are connected with their usage. Use of lethal pesticides, inability to predict weather, dubious quality of resultant composites, practicing of treatments causing pollution, and utilization of agricultural land at the cost of crops are some of these aspects [25].

Using natural materials to produce fiber reinforcements used in cosmetics also works against globalized manufacturing system of automotive industry. To reduce cost of the product, it is advisable to use local raw material, which unfortunately, is not possible every time as production of crops depends mainly on the climatic conditions which vary in different parts of the world. For instance, broom has three yields per year in the Mediterranean countries, whereas Germany is popular for its growth of sisal. Similarly, Brazil uses coconut fiber in production of foams for seats, whereas, India uses abaca fibers for this purpose.

Considering this situation, it is difficult to produce similar products or modules in different countries by utilizing different locally grown crops. Sole alternative is, to export the raw material which of course, is a costly procedure.

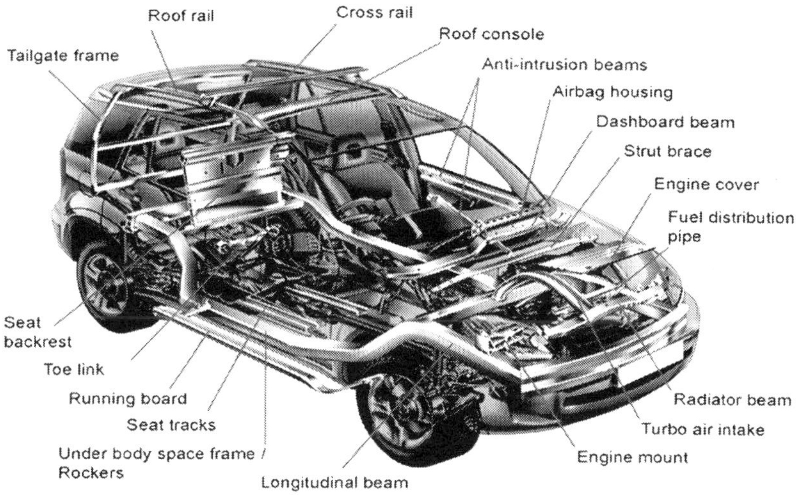


**Figure 28:** Future uses of composite materials in automotive parts

Carbon fibers seem to have bright future in automotive industry. They are now used in many applications, chiefly as the main structural material of high performance autos. However, variations in the cost of carbon fibers and difficulty in their availability remains a major issue to date.

A new class of composite materials is evolving. Founded on Nano-technologies like carbon nanotubes, they are added to conventional composite materials to improve their mechanical characteristics. Some of the additional advantages include altered penetration of fluids and gases, reduction in flammability, improvement in thermal characteristics, enhanced thermal and electrical conductivity. Two exciting examples of this application are the Chevrolet Astro 2002 and the GMC Safari. Polypropylene composite has been used in both applications after being reinforced with clay nanoparticles. Even with only 2.5 % nano-filler, the stiffness of components is comparable to their counterparts containing then times the quantity of talc filler. Reason is the bigger surface areas of nanoparticles.

Car manufacturers are always looking for new materials to play with; together with composite material suppliers they are ever ready to explore the opportunities in this field and thus can be relied upon for significant developments of new materials to be used in automotive applications.

## REFERENCES


1. Allison, J. E., & Cole, G. S. (1993). Metal-matrix composites in the automotive industry: opportunities and challenges. JOM Journal of the Minerals, Metals and Materials Society, 45(1), 19-24.
2. Ashby, M. F., & Brechet, Y. J. M. (2003). Designing hybrid materials. Acta materialia, 51(19), 5801-5821.
3. Baillie, C. (Ed.). (2005). Green composites: polymer composites and the environment. CRC Press.
4. Carruthers, J. J., Kettle, A. P., & Robinson, A. M. (1998). Energy absorption capability and crashworthiness of composite material structures: a review. Applied Mechanics Reviews, 51(10), 635-649.
5. Chapple, S., & Anandjiwala, R. (2010). Flammability of natural fiber-reinforced composites and strategies for fire retardancy: a review. Journal of Thermoplastic Composite Materials, 23(6), 871-893.
6. Friedrich, K., & Almajid, A. A. (2013). Manufacturing aspects of advanced polymer composites for automotive applications. Applied Composite Materials, 20(2), 107-128.
7. Hashim, J., Looney, L., & Hashmi, M. S. J. (1999). Metal matrix composites: production by the stir casting method. Journal of Materials Processing Technology, 92, 1-7.
8. Holbery, J., & Houston, D. (2006). Natural-fiber-reinforced polymer composites in automotive applications. JOM Journal of the Minerals, Metals and Materials Society, 58(11), 80-86.).
9. Jambor, A., & Beyer, M. (1997). New cars—new materials. Materials & design, 18(4), 203-209.
10. Jones, R. M. (1998). Mechanics of composite materials. CRC press.
11. Kurtaran, H., Eskandarian, A., Marzougui, D., & Bedewi, N. E. (2002). Crashworthiness design optimization using successive response surface approximations. Computational mechanics, 29(4),

409-421.

12. Mangino, E., Carruthers, J., & Pitarresi, G. (2007). The future use of structural composite materials in the automotive industry. International Journal of vehicle design, 44(3-4), 211-232.
13. Potter, K. (2012). Resin transfer moulding. Springer Science & Business Media.
14. Simacek, P., Advani, S. G., & Iobst, S. A. (2008). Modeling flow in compression resin transfer molding for manufacturing of complex lightweight high-performance automotive parts. Journal of composite materials.
15. Thornton, P. H., & Jeryan, R. A. (1988). Crash energy management in composite automotive structures. International Journal of Impact Engineering, 7(2), 167-180.
16. Yang, Y., Boom, R., Irion, B., van Heerden, D. J., Kuiper, P., & de Wit, H. (2012). Recycling of composite materials. Chemical Engineering and Processing: Process Intensification, 51, 53-68.

# CHAPTER 3

# DOMESTIC APPLICATIONS OF COMPOSITE MATERIALS

## INTRODUCTION

Extensive variety of raw materials, for example, pineapple leaves, coconut husks, banana stems, oil palm biomass, rice straw, bamboo, rattan, and wood squanders are profoundly accessible around the globe and have been utilized to produce Biocomposites for furniture applications. The mechanical, physical and chemical properties of these inexhaustible assets are like those of wood, and they might be appropriate raw materials for wood based boards (Khalil et al., 2010a). Both local wood markets and global furniture makers are shifting towards environment-friendly products. Examples of Biocomposites materials are particleboard; furniture makers utilize wood plastic composites (WPC), plywood, medium density fiberboard (MDF), laminated veneer lumber (LVL) and hardboard for a broad range of applications. Presently, oil palm biomass is substitute material for wood-based products. Moreover, many manufacturers are making board items from oil palm biomass (Khalil et al., 2010c). Former research conducted on, oil palm trunks (OPT), oil palm fronds (OPF), oil palm press fibers (PPF) and oil palm empty fruit bunches (EFB) has revealed that these raw materials are appropriate for producing new, safer, cheaper, sustainable, and high-value goods using innovative techniques. Biocomposites such as WPC, LVL, OSB, MDF, agro-lumber, and plywood can be produced commercially for their use in furniture development (Yusoff &Koh, 2001).

The mechanized furniture production from wood was developed during the Renaissance in the first mid of the 15$^{th}$ century. People who purchased these products used this furniture as a mark of their politics, wealth, position, power and origins (Pawlak, 2007). Moreover, there is no change in material selection since that time, because the solid wood had excellent performance, an elegant appearance, simple manufacturing and accessibility, and resources were easily available.

Furthermore, important features for the selection of wood materials in the 15$^{th}$ century included their unique and authentic surface textures, trustiness, and high perceived value. Solid woods were used as raw material early international exhibits, and it continued to grow since the beginning of the 1870s. At that time, solid woods were readily available, and they were inexpensive. Since 1980, furniture production industries in the Association of Southeast Asian Nations (ASEAN) have grown rapidly. ASEAN countries are producing over 70 to 80% of all wood-based furniture worldwide. The economy of ASEAN countries has been enhanced because of the increased demand of rubberwood furniture which was mainly focusing on the agricultural sector. To date, the primary material used for the furniture making is wood. However, with developments in science and technology, the materials used in furniture making have experienced several changes. Mixtures of constituents in both wood and other raw materials from biomass are available as an alternate for pure wood materials to enhance availability and increase productivity and a recent study reported new raw materials, such as the combination of rubber with wood oil palm EFB, for furniture parts. (Khalil *et al.* 2010c).

The present worldwide demand for furniture products remains high despite the several economic downturns. Vietnam and China establish strong competitors for Indonesia, Thailand, and Malaysia, because of their original designs, lower manufacturing costs, the good work ethic of their labor force and their aesthetics. The part of the government is also critical for developing industries, providing status for tax releases and investment tax grants that permit businesses in the industry to become more readily recognized. A high-quality product, business-friendly environment, and a large potential market are important for furniture industry to surpass expectations and continue to grow at very high rate. (Puasa *et al.,* 2010; Bovea & Vidal, 2004a). A large number of articles are available that describe the findings on composite materials and bio-composites with fascinating results. (Khalil *et al.*, 2011; Khalil *et al.*, 2010b; Khalil *et al.*, 2010c; Bhat *et al.*, 2010; Yusoff & Koh, 2001).

Design features of Biocomposites and their production concerning novel furniture design and their demand based on population growth have not been properly considered till date. This chapter contains, a short introduction to oil palm biomass and their uses in Biocomposites, furniture, and design. Different types of oil palm Biocomposites *vs.* other materials, and the challenges faced by oil palm Biocomposites regarding innovative design and growing populations are also discussed in this chapter. Till now, no such review was available which cover all these aspects of the furniture industry. This chapter, therefore, is proposed to join all the accumulated knowledge and discuss some relevant findings and the trends in this subject area.

## WOOD PRODUCTS BY CATEGORY

### Oil Palm Biomass

The industry of oil palm has been growing at a remarkable rate over the 1$^{st}$ decade of 21$^{st}$ century. Thailand, Malaysia, Indonesia, and other Asian countries are mainly considered to be agricultural countries; even Malaysia is also famous for its other plantations, such as cocoa coconut and rubber. The use of oil palm can differ due to its use in recycling activities which uses about 90% of the oil palm biomass. (Khalil & Rozman, 2004). Due to various initiative promoting the production of enterprises, the level of use of oil palm is expected to increase further. Additionally, the raw material from the industry of oil palm has other uses as well which includes the production of plywood paper, pulp, and other Biocomposites. Presently, the oil palm production in Malaysia is 50% more than other countries producing oil palm; e.g. only 30% of the total oil palm produced by competing countries is provided by Indonesia. The oil palm production in research programs is more than double of the wild plants in Africa. Due to this growth in production of oil palm Malaysia has replaced Nigeria as the world's leading exporter and producer of oil palm (Khalil & Rozman, 2004). Statics show that residual oil palm biomass of about 40 million tons wastes from oil palm trees such as OPT, EFB, PPF, OPF are produced per year in Malaysia (Baharuddin et al., 2009). After reviewing the current research in modern furniture raw materials, it was observed that the total oil palm biomass used is only 10%, and the remaining is underutilized. (Khalil and Rozman 2004).

The alternatives to wood materials such as cellulosic from both OPF and coconut trunks can be used as raw materials. (Khalil & Rozman, 2004; Amouzgar *et al.,* 2010). Additionally, the furniture industry might be affected by the wood products produced from OPT. OPT has been processed as alternative materials for making wood furniture because it can be compared with rubberwood (RW) in quality and to other solid wood species. Moreover, OPT can be considered as an assist for furniture industries that entirely rely upon RW, which can change nation's ecosystem. OPT lumber was recognized as an alternative material for rubber-wood (RW) because of its mechanical and physical properties, high resistance to pests and thermal stability. These factors are important for making furniture because they ensure resilience, strength, durability and quality of the furniture allowing the customer to use the product for a long period. (Bhat *et al.,* 2010).

The value of the furniture products can be increased by certain key factors such as design and effect on the atmosphere, which according to recent work is critical to lead the market demand (Bovea & Vidal, 2004a). The use of increased value of OPTL has been highly encouraged and praised because they have increased innovation and technology and low environmental impact. The main components of the oil palm biomass are shown in Fig. 29, which shows both the process flow to the final manufactured products and the process of using oil palm waste as oil palm biomass. The method is comparable to the process that involves fronds and palm trucks except that waste is a mixture of fresh fruits and fruit bunches to manufacture fruit fibers.

Every part of oil palm trees is useful unlike other wood alternatives, which is necessary for environment sustainability. Research in producing friendly bio-based products are advancing worldwide.

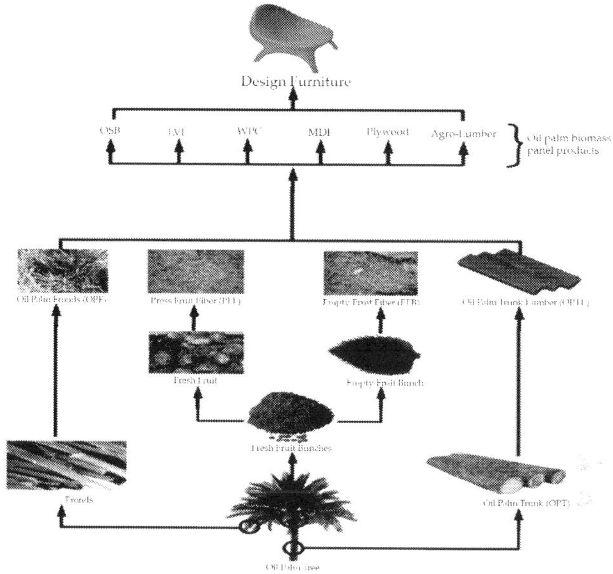


**Figure 29:** Process flow of oil palm biomass materials (OPF, PFF, EFB, and OPTL)

## Agro-Based Biocomposites Products

In some countries, there is an abundance of forests that produces wood and timber. Hence, the use of wood waste is very cost-effective. The processing of waste wood in industries contributes to the production of raw material furniture production demand (Bovea & Vidal, 2004a). The amount of wood waste that can be processed into the final product is only 50%, and economy might be harmed excess wastes. Composite wood goods are used in both structural and non-structural applications, which includes paneling for both external and internal furniture and also for construction Maloney (1986) presented a classification system to classify a range of wood based composites (Stark *et al.,* 2010; Maloney, 1986).

This classification scheme is shown in Table 2, which defines the types of wood panel products that reflect the latest advancement in agro-based Biocomposites.

**Table 2:** Grouping of Wood-Based Composite Materials (Maloney, 1986)

| | |
|---|---|
| Laminates | Plywood<br>Laminated veneer lumber (LVL) |
| Veneer-based material | Wood-non-wood composite |
| Composite material | Cellulosic fiberboard<br>Particleboard<br>Waferboard<br>Flakeboard<br>Hardboard<br>Oriented strand board (OSB) |
| Edge-adhesive-bonded material | Wood fiber-agricultural fiber composite Wood fiber-plastic composite |
| Wood-non-wood composite | Lumber panel |

Production of Biocomposites from agricultural wastes has been increasing per year from 2004 to 2011, and it accounts for half of the market wood-based production. Biocomposites products demand has been rising per year due to the efforts of producers and enhanced customer awareness of the sustainability and importance (Leao *et al.,* 2010). Table 3 illustrates the subsequent development of Biocomposites from agro materials. Researchers are studying the potential of these composites to determine if they have advantages.

An effective experiment manufactured Biocomposites furniture using the raw material of fiber chips. The advantage of using particleboard made from fiber chips is that it has better machining parameters and enhance the efficiency of the raw material usage (Bhat *et al.,* 2010). Novel research of combining both OPT and EFB manufactured a new blend of plywood with enhanced quality relative to current plywood, particularly about screw withdrawal, shear strength and flexural strength (Salleh *et al.,* 2011; Khalil *et al.,* 2010c; Khalil & Rozman, 2004). Example include studies using oil palm trunks to make molded laminated veneer oil palm for furniture showed that melded veneer oil palm performed better than normal wood veneer, and was appropriate for making gainful furniture products Still, further research is currently going on to add fibers to biocomposites products, and use of composites in furniture production is essential for enhancing the durability and quality of MLVOP (Leao *et al.,* 2010).

**Table 3:** Chronological arrangement of events in the Development of Agro-based Bio-composites

| Agro-Based Biocomposites | Year | References |
|---|---|---|
| Plywood | 1865 | (Sheldon Shi 2006) |
| Oriented Strand Board (OSB) | 1970 | (Cai &Winandy, 2005; Shi & Walker, 2006) |
| Laminate Veneer Lumber (LVL) | 1944 | (Kairi, 2005) |
| Medium Density Fiberboard (MDF) | 1970 | (Khalil &Bhat, 2010) |
| Particleboard | 1975 | (Takahashi, 2001) |
| Hardboard | 1926 | (Thomson, 2003) |
| Wood Plastic Composites (WPC) | 1970 | (Pritchard, 2004) |
| Agro-Lumber | 2010 | (Khalil *et al.*, 2010a) |

The demand for household furniture has been increased rapidly than the demand since 1997 for any other wood-based product because of yearly world's population growth. Besides, consumers need long-lasting, high-quality furniture (Levesque 2000). The growing world's population is the main factor in managing the natural resources. The highest population in South Asia lies in Indonesia and is opposed with the problems of giving home in safe and friendly environment despite its large area if fertile land. Management of natural resources instead of selling them on the world market is their Biocomposites or raw form is necessary for improving the quality of life (Rajan and Amin 2001).

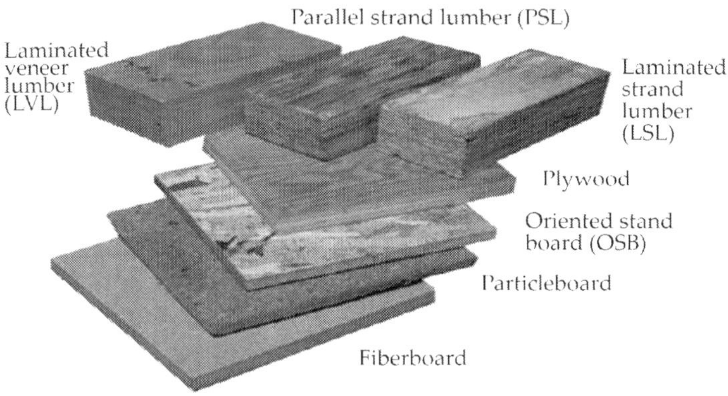


**Figure 30:** Different bio-composite products

## Oriented Strand Board (OSB)

Oriented strand board (OSB), is an engineered structural material produced from thin wood strands joined together with a water resistant resin using pressure and heat. Oriented Strand Board technology and the raw material used basically evolved from wafer-board technology, which mainly used aspen the late 1970s. . As the company learned to control their placement, orientation and strand size, and the durability, utility performance of OSB products enhanced to the point where they performed at a quality similar to structural plywood (Kong, 2001). As a result, oriented strand board has become accepted and its usage has extended to substitute softwood plywood in construction products worldwide. Oriented strand board differs from wafer-boards in a way that oriented strand board wood strands are not randomly placed but oriented. Oriented strand board was easily accepted since debuting in the market in 1970 (Youngquist, 1999).

In many regions of North America oriented strand board is used for structural and residential constructions, and both Canada and US allow oriented strand board panels to be used for the same cause as plywood on thickness basis. Wood fibers, degree of strand orientation in the surface layer and interweaving of long strand of wafers are mainly responsible for the strength of OSB. Composite wood products such as OSB, MDF, Hardboard and particleboard are manufactured by Diphenylmethane di-isocyanate (MDI) mixture (Youngquist, 1999).

## Laminated Veneer Lumber (LVL)

Laminated Veneer Lumber (LVL) is a wood composite that has been manufactured by using adhesive such as phenol formaldehyde, urea formaldehyde, phenol resorcinol formaldehyde and melamine urea formaldehyde (Shi &Walker, 2006). Laminated Veneer Lumber is a technique that joins together parallel veneer sheets and is a new and novel innovation in the production industry. This technique shows that joining veneers can enhance the strength of Laminated Veneer Lumber that only uses veneers. The huge demand for Laminated Veneer Lumber materials shows that Biocomposites are comparable with the existing solid wood products. Because of research and development activities Laminated Veneer Lumber based Biocomposites have entered the global market. The LVL materials have advantages over ordinary plywood which include their formation of possessing perfect characteristic combinations combined with quality assurance and high tensile strength. In US, aircraft components have been manufactured using LVLs since 1944.

Innovative studies have been conducted on Laminated Veneer Lumber moulds using high frequency technology to form products into various shapes which depends on the mould design. Natural fibers have both mechanical and physical properties that can be used to manufacture high quality composite by adding veneers with resins (Leao et al., 2010). Additionally, the investigation of potential of timber used to manufacture Laminated Veneer Lumber veneers that had one or more layers was done. The Laminated Veneer Lumber manufacturing methods enable the production of parts with about any dimension; the length of LVL can range up to 25 m and thickness up to 19 to 75 mm are possible without restrictions and are subject to transportation and operating machinery (Shi and Walker 2006).

## Wood Plastic Composites (WPC)

Wood plastic composites are made of plastics and wood fibers that comprise of several non-organic filler materials and lignocellulosic fibers. Wood plastic composite are also called natural fiber plastic composites. The origin of WPC manufacturing began in the US many decades ago (1970) and sustained in Japan and Europe. Wood plastic composite industry has been growing each year since 1970s (Pritchard, 2004). This growth is due to the fact that producers mainly produce Wood plastic composite by mixing the other materials. Statistical data shows that Wood plastic composite industry only covered a portion of total wood produced for the industry at 1970. However Wood plastic composite has become an important product for the production sector during the changes of world toward the present use of important Biocomposites sources (Smith, 2002).

Now a days, Wood plastic composite are a famous replacement to natural wood Biocomposites and are thought as novel and quality products (Rowell, 2005). The huge demand for wood materials as the durability and design motifs of wood improved by emerging technologies have made WPC to attract global market. WPS belong to new generation of Biocomposites that are durable, more eco-efficient, environment friendly and dynamic. Research and development scientists are developing WPC that have enhanced properties suitable for use in commercial production of end products, especially in automobiles, building construction and furniture design (Rosli, 2010). The strength of Wood plastic composite is about eight times greater than ordinary wood; however it can be recycled or reprocessed into a biodegraded or new product. The components of WPS maintain the standards of materials to substitute the existing thermoplastic and wood containing materials, which are widely used to make products. Hence, it is not astonishing that WPS are popular in both North America and Europe; they are supposed to be most important, highest quality available and protective for environment.

## Medium Density Fibreboard (MDF)

Fiberboards include High density fiberboard, medium density fiberboard and low density fiberboard and each one of these is made by both dry and wet processes. Composites panels such as plywood, particle board and MDF are mostly used in production of products such as flooring, furniture and doors because they have low cost and production is easy to produce (Khalil & Bhat, 2010; Nemli, 2000). Medium density fiberboard is a wood-based composite which contains panel composed of mixture of synthetic resins such as Urea formaldehyde and lignocellulosic compounds. Two main sources for the production of MDF have been identified: rubber-wood (RW) and oil palm EFB pulp fibers. However recent studies have shown that rubberwood have inferior than oil palm

EFB pulp fibers because later contain fibers with better buffering capabilities and pH value. Medium density fiberboards made from above suggested materials are appropriate to substitute those produced from dust that are known as less resistant (Nadlin, 2009). Initially, MDFs were coated with layers made from mixture of unmaintainable materials, such as lacquer, veneer, paper, PVC foils, laminating plastics and melamine paper. These layers do not have same characteristics of layers, because both the plastics and chemical mixtures are used to increase the strength of wood and to smooth its surface.

Mostly researchers use EFB as the main raw material over the presently available resources for production of MDFs. The material and processes used in production of MDF greatly affect the stability and mechanical properties of wood (Khalil *et al.,* 2010b). Medium density fiberboard have been used as a replacement to wood in the particle industry because of increasing deforestation. The annual rising production of MDF results from the shortage of natural resources and as well as offers an opportunity for manufacturer to use sustainable resource materials for production of their products. The color, texture, uniformity of rubber-wood fibers in MFD makes it ideal. The medial compression of MDF is highly valued at price relative to plywood, hardboard, particleboard (Thomas, 1997).

## Particleboard

The rise in agricultural areas, huge demand for wooden materials and forest fires have accelerated the importance of composites particleboard than solid wood. Presently, particleboard provides the customers constant design flexibility and quality needed for the efficient production of products. The demand for particleboard which is an alternative raw material was affected by its low cost as compared to traditional solid wood. The benefits of using particleboard are not restricted to processing a constant quality for the design of product but also use of agricultural wastes in the production process. Particleboard is a non-structural product produced from the by-products of wood particle from the production of other wooden products. Synthetic binders or resins are mixed with these wood particles and a panel is formed under pressure and heat. The board is made up of many layers: two outside layers and a core. Particle board is usually used to manufacture products appropriate for board densities, specific geometric resin levels and manufacturing processes. Additives can be added during the production to improve its certain properties such as increasing both its moisture and fire resistance and improving dimensional stability (Khalil & Hashim, 2004). Grading of the produced particleboard is done on its quality and particles are regularly tested to make sure that it is certain to fulfil the strict industry performance standards. Products such as shelves, floor underlayment, kitchen cabinets, office furniture and many other products based on wood can be replaced with particleboard. The lignocellulosic fibrous behavior is unique to particleboard relative to other types of boards, which are made from many lignocellulosic fibers such as jute and kenaf, which enhances the low resilience sound absorption of particleboard (Shi &Walker, 2006; Youngquist, 1999).

## RUBBERWOOD *VS.* OTHER MATERIAL BASED BIO-COMPOSITES FOR FURNITURE

Since the ancient times, the mankind has been utilizing tree-wood for many purposes. Wood has been an essential part of the human civilizations, and it has been widely used for fuel, transportation, furniture, weapons and many other household items. Moreover, wood was also utilized for communication purposes by tapping two wooden-pieces each other. It would have been impossible for human race to survive the harsh climates without the use of tree-woods.

Rubber-wood (RW) tree is considered to be a primary source of wooden raw materials because of its widespread use in furniture industry of Thailand, Indonesia, Malaysia, and different other regions wood producing countries. Due to unique eco-friendly nature of RW, it is mostly preferred as a viable material for production of household items. Studies have shown that consumers prefer to buy the products made from sawn timber over other wooden products because of many unique features of RW such as, high durability, easy processing, and high aesthetic value. Rubber-wood has a light straw or light brown colour with a density varying from 550 kg/m3to 650 kg/m3. The trunk length of RW is around 3-5 m and its diameter ranges between 30 cm and 35 cm (Khalil et al., 2010c). The demand for fresh wood has increased significantly over the years. In the past, the wood-demand was insignificant as compared to the availability of the forests. However, due to immense increase in the population of the world, timber demand has increased considerably, ultimately consuming the forests. The current trends in the demand of wood have endangered the ecosystem, due to rapid deforestation, for satisfying human wood requirements (Yusoff & Koh, 2001). There has been a drastic decline in the acreage of RW (2 million hectares to 1.6 million hectors from 1995 to 2008) which has resulted in the shortage of RW in timber markets across the world. Now a days, furniture manufacturing from wooden raw materials has become quite expensive due to the scarcity of the raw wood (Roger, 2000). Malaysia, Thailand, Indonesia, China, India, Vietnam, and a few other countries are famous for providing top-quality wooden products manufactured primarily from rubber-wood. The US, West Asia and Japan are major importers of furniture from Malaysia. Moreover, European countries have also started to import furniture products from Malaysia which is unfolding new horizons for Malaysian wood markets.

Incorporation of novel design and technologies are essential to compete with the international wood markets. Modern technologies can assist in bulk production of furniture products, which can be exported to Europe and other countries. For instance, Malaysian government formulated 2nd Master Plan (1996-2005), which recognized furniture industry as a target market. Later on, the National Timber Industry Policy (NATIP) was established to control the direction of the country's wood industry from 2009 to 2020. The formation of NATIP was essential for maintaining a competitive and sustainable environment for Malaysian timber market in a challenging international market. Other countries such as UK, USA, Sweden, Holland etc. have also taken such initiatives to uplift their industries.

Rattan and bamboo are versatile materials, which can replace rubber-wood for furniture production, because of their low-cost, light weight, low cost, and environment friendly nature (Bansal & Zoolagud, 2002). Most often, rattan is mistaken for bamboo because of their identical nature and origin, i.e. Both species are derived from the same plant kind. However, these two species differ in colour, size and some physical properties. Bamboo grows faster and it can stretch to a height of about 30 feet in around 2 years. Due to the convenience in the planation of bamboo, abundant sources of bamboo wood are available for production of furniture (Saville, 2008). Bamboo has a smooth surface with various aesthetic variations based on its colour, i.e. yellowish, brown, and green. Besides, substantial mechanical strength, the unique appearance of bamboo and its excellent durability makes it an ideal material for furniture manufacturing (Bansal & Zoolagud, 2002).

## FURNITURE DEMANDS OF INCREASING POPULATIONS

The requirement of household amenities is closely associated to the density of the population for any specific country. Increment in market demands results in shortage of the raw materials (Ljungberg, 2007). Currently, emerging furniture markets are utilizing Biocomposites as a substitute for the wood to minimize deforestation and its impact on the environment (Larsson-Brelid et al., 2010). According to the surveys, the world's population was about 7 billion in 2011, and it will continue to grow steadily depending on the improvements in health and the environment sector (Rohana, 1999; Bureau, 2011). The extrapolated world population graph is shown in Fig.31. It is estimated that the world population will increase to 10 billion in next 40 years (after 2050). The increase in the world population will ultimately result in the increased furniture demand for domestic use. Consequently, there will be more deforestation to satisfy the furniture requirements of the population (Bureau, 2011).

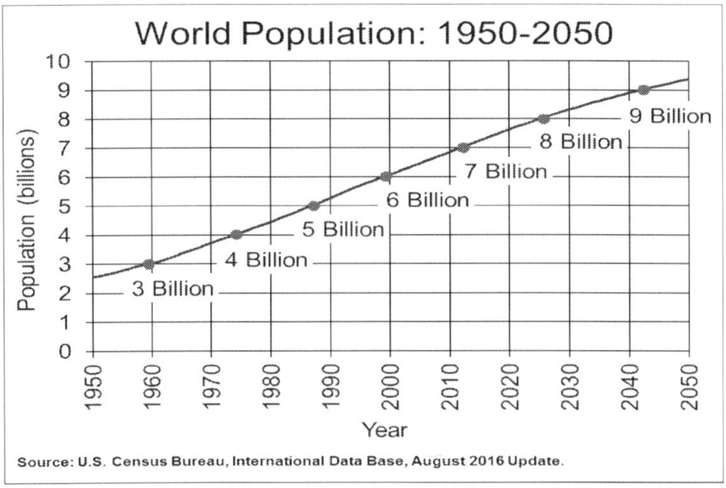


**Figure 31:** World population data (2016)

Product demand is not necessarily dependent on the population growth. There are some countries which have lower pollution but abundant timber resources. Indonesia is one of the low population countries with fertile forests. The current research on alternative wood products can prove expedient for Indonesia to address the issue of deforestation. Deforestation has caused numerous problems for most of the countries. These countries can utilize the research results by shifting their furniture products on alternative resources. Presently, many wood producing countries have a chance to produce and export the Biocomposites based furniture products at a bulk scale. There is a range of raw materials for manufacturing of Biocomposites based furniture. Deliberate comparison of the products can result in cost effective and efficient furniture products.

## FURNITURE AND DESIGN

Furniture is a common commodity found in every house in different forms. The choice of the furniture product typically depends on various features such as style, size, shape, shading and material type (Shukri et al., 2011). Modern furniture design encompasses a wide variety of environments such as offices, homes, hotels, restaurants, shopping centers, public parks, hospitals and bus stops (Alli & Rahman, 2008). Different varieties of the furniture are categorized on the basis of their design and functionality. Today, multifunctional furniture has become a general trend in conjunction with the modern lifestyle. Multifunctional furniture is designed to be eco-friendly and smart product. In recent years, the furniture industry has shifted to modern manufacturing techniques for production of cost competitive furniture (Reis & Wiedemann, 2010). Scientific developments have enabled the furniture industries to produce the goods at faster and efficient rates. High rates of furniture production also affect the economy in a positive way (Saville, 2008). Malaysia is one of the biggest producers of furniture products in the world. Records show that that Malaysia earned RM5.4 billion from timber-based products in 2004. It is obvious that Malaysian furniture industry has grown steadily to become the global giant in furniture production (Nadlin, 2009).

## CHALLENGES FOR THE UTILIZATION OF AGRO-BASED BIOCOMPOSITES FOR FURNITURE APPLICATIONS

### Industry Structure

The modern furniture industry of any country relies on three kinds of wood reserves such as government reserves, freehold lands and the permanent forest reserves. Malaysian timber production showed a decline in 2010 when numerous forest lands was reserved for development projects. Considering the issues of deforestation, the Ministry of Plantation Industries and Commodities (MPIC) established timberland farms to cater the problem of low timber production. There are some countries which prohibit the export of timber products to minimize the effect of furniture production on the environment. The production of sustainable raw materials has become a challenge for the timber industry. There are plentiful agricultural products such as empty fruit bunch (EFB), oil palm frond (OPF) and oil palm trunk (OPT), that are not presently being utilized. Despite being considered as waste materials, these agriculture products can prove to be excellent substitutes for the wood industry.

**Figure 32:** Novel design from agro-based Biocomposites, (a) Wood plastic composite, outdoor furniture (b) kiwi coffee table, (c) Furniture from Particle-board (d) Hardboard (e) LVL (f) OSB Table (g) Chair (h) Spring seat design

## Sustainable Supply of Raw Materials and Human Resource Development

To sustain the growth of the wood industry, numerous steps needs to be taken to for a consistent supply of raw materials for furniture production.

Considering the prevalent challenges of the furniture industry, the Malaysian government has strictly instigated numerous initiatives to discover substitute materials from other resources to upkeep the Malaysian timber industry. One of the alternative raw materials for timber industry is agricultural residue. Agriculture wastes are widely available materials which can be used to support the timber industry. The Biocomposites production is still new in Southeast Asia, and the shift of a country's industry from conventional to alternative materials is likely to diminish the local workforce supply. The Malaysian government has attempted to back the Biocomposites industry by instituting a number of wood training centers including WISDEC and FITEC to enhance the production capacity and worker skills for efficient manufacturing of the furniture products (Oosterlaken, 2009).

## Innovation and Technology

Incorporation of modern technology in the manufacturing industry is essential to cope with the competitive furniture markets. Use of technology ensures the consistent production of high quality products at a steady rate. The efficiency of a machine depends on its manufacturing speed and its ability to produce durable products consistently. It is imperative for the modern manufacturing techniques to ensure the efficient and reliable production of the furniture.

## REFERENCES

1. Abdul Khalil, H. P. S., Nurul Fazita, M. R., Jawaid, M., Bhat, A. H., and Abdullah, C. K. (2011). Empty fruit bunches as a reinforcement in laminated bio-composites. Journal of Composite Materials 45(2), 219-236.
2. Abdul Khalil, H. P. S., and Rozman, D. (2004). Gentian dan Komposit Lignoselulosik. Universiti Sains Malaysia Publisher, Penang, Malaysia.
3. Abdul Khalil, H. P. S., and Hashim, R. (2004). Komposit Panel Berasaskan Sumber Kayu. Universiti Sains Malaysia Publisher, Penang, Malaysia.
4. Abdul Khalil, H. P. S., and Bhat, A. H. (2010a). Oil palm biomass: Fibre cultivation, production and its varied applications. In: Oil

Palm: Cultivation, Production and Dietary Components, S. A. Penna (ed.), Nova Science Publisher, Inc., 21-34.

5. Abdul Khalil, H. P. S., Firdaus, M. Y. N., Jawaid, M., Anis, M., Ridzuan, R., and
6. Mohamed, A. R. (2010b). Development and material properties of new hybrid medium density fibreboard from empty fruit bunch and rubberwood. Materials and Design 31(9), 4229-4236.
7. Abdul Khalil, H. P. S., Nurul Fazita, M. R., Bhat, A. H., Jawaid, M., and Nik Fuad, N. A. (2010c). Development and material properties of new hybrid plywood from oil palm biomass. Materials and Design 31(1), 417-424.
8. Akihiro Takahashi, a. E. A. (2001). Application of MDI binder towards environmental friendly wood-based industry. Ceminar on Wood-Based Panel Products, Forest Research Institute Malaysia, Kuala Lumpur, 6.
9. Alli, H. H., and Rahman, A. R. A. (2008). Reka Bentuk Perindustrian: Pengenalan. Dewan Bahasa dan Pustaka Kuala Lumpur.
10. Amouzgar, P., Abdul Khalil, H. P. S., Salamatinia, B., Zuhairi Abdullah, A., and Issam, M. (2010). Optimization of bioresource material from oil palm trunk core drying using microwave radiation; A response surface methodology application. Bioresource Technology 101(21), 8396-8401.
11. Anon (2009). The most unusual lounge chairs in the world. http://naturalinteriordesign1.blogspot.com/2009/09/15-most-unusual-lounge-chairsin-world.html
12. Anon (2011). World Population Data Sheet. <http://www.prb.org/pdf11/2011population-data-sheet_eng.pdf>.
13. Baharuddin, A. S., Kazunori, N., Abd-Aziz, S., Tabatabaei, M., Abdul Rahman, N. A., Hassan, M. A., Wakisaka, M., Sakai, K., and Shirai, Y. (2009). Characteristics and microbial succession in co-composting of oil palm empty fruit bunch and partially treated palm oil mill effluent. The Open Biotechnology Journal 3, 92-100.
14. Bansal, A. K., and Zoolagud, S. S. (2002). Bamboo composites: Material of the future. Journal of Bamboo and Rattan 1(2), 119-130.
15. Bovea, M. D., and Vidal, R. (2004a). Increasing product value

by intergrating environment impact, cost and customer valuation. Resources Conservating & Recycling 41, 12.

16. Bovea, M. D., and Vidal, R. (2004b). Materials selection for sustainable product design: a case study of wood based furniture eco-design. Materials & Design 25, 5.
17. Brower, C., Mallory, R., and Ohlman, Z. (2009). Experimental EcoDesign: Product, Architecture, Fashion. RotoVision SA, UK.
18. Bureau, P. R. (2011). World Population Data Sheet 2011. http://www.prb.org/Publications/Datasheets/2011/world-population-data-sheet.aspx
19. Cai, Z., and Winandy, J. E. (2005). Opportunity and development of bio-based composites. Proc., Proceeding of International Workshop on Prefacricated Bamboo Panel Module, Beijing, China, 11.
20. Jaime (2010). Chris Rucker at Industrial Gallery. http://design-milk.com/chris-ruckerat-industry-gallery/
21. John, M. J., and Thomas, S. (2008). Biofibres and biocomposites. Carbohydrate Polymers 71(3), 343-364.
22. Kairi, M. (2005). Interaction of R&D and business development in the wood products industry, Case Kerto®–Laminated Veneer Lumber (LVL). Helsinki University of Technology Laboratory of Wood Technology Reports, Helsinki University of Technology Department, Finland, 237.
23. Kong, H. W. (2001). Oriented Starnd Board (OSB) - A new engineered wood product for Malaysia. Seminar on Wood Based Panel Products, Forest Research Institute Malaysia, Auditorium, Forest Research Institute Malaysia, 4.
24. Larsson-Brelid, P., Walinder, M. E. P., Westin, M., and Rowell, R. (2010). Ecobuild a center for development of fully biobased material systems and furniture applications. Molecular Crystals and Liquid Crystals 484, 8.
25. Leao, A. L., Souza, S. F., Cherian, B. M., Frollini, E., Thomas, S., Pothan, L. A., and Kottaisamy, M. (2010). Agro-based biocomposites for industrial applications. Molecular Crystals and Liquid Crystals 522, 18-27.
26. Levesque, Y., Lumber Manufacturing Department, Eastern Division

(2000). Lumber and Value-Added Wood Products: Special Report. Forintek canada Corp., Canada.

27. Liansheng, Y., Maili, S., Tao, W., and Wu, Z. (2002). The pretreatment of carbon fibres for 3D C/SiC composites. Science and Engineering of Composite Materials 10(1), 55-58.
28. Ljungberg, L. Y. (2007). Materials selection and design for development of sustainable products. Materials & Design 28(2007), 466-479.
29. Maloney, T. M. (1986). Terminology and products definitions—A suggested approach to uniformity worldwide. Proc., 18th International Union of Forest Research Organization World Congress, Ljubljana, Yugoslavia.
30. MIFF (2011). Malaysian International Furniture Fair (MIFF). http://www.versalink.com/news/events/malaysia-international-furniture-fair-miff-2011
31. Mindmarsh, M. (2011). Custom designed furniture for NMIT. http://urbismagazine.com/articles/furniture-designed-specifically-for-nmit/
32. Nadlin, R. S. (2009). Jualan Papan Gentian Sawit Disasarkan RM20j. Utusan Malaysia, Utusan Malaysia, Kuala Lumpur.
33. Rajan, S., and Amin, M. N. Z. M. (2001). International standards development activities for wood-based panel products. Ceminar Wood Based Panel Products, Forest Research Institute Malaysia, Auditorium, Forest Research Institute Malaysia, 8.
34. Reis, D., and Wiedemann, J. (2010). Product Design In The Sustainable Era. Taschen, Germany.
35. Rohana, M. (1999). Papan Gentian Daripada Sisa Kelapa Sawit. Utusan Malaysia Kuala Lumpur, 1.
36. Roger, T. (2000). A review of adhesives for furniture. International Journal of Adhesion and Adhesives 20(4), 269-272.
37. Rosli, M. Y. (2010). Pilot scale feasibility study for the production of wood plastic composite (WPC) Pellet from sawdust for furniture and automotive industry. Conference on Innovation in Biocomposite and Paper, Malaysia Timber Industry Board (MTIB), Kuala Lumpur, Malaysia.

38. Rowell, R. M. (2005). Moisture properties, Handbook of Wood Chemistry and Wood Composite. R. M. Rowell (ed.), Forest Products Laboratory, Wisconsin, Madison, WI 77-98.
39. Salleh, A. H., Kamal, I., Rashid, N. A. A., Abdul Khalil, H. P. S., Ahmad Shakri, M. S., and Mahmud, S. Z. (2011). Preliminary study of moulded laminated veneer oil palm (MLVOP). Modern Applied Science 5(4), 11-18.
40. Samarreda (2011). Recycled wood chairs from particleboard by Samarreda. http://www.trendir.com/ultra-modern/recycled-wood-chairs-from-particleboard-bysamarreda.html
41. Saville, S. (2008). Design Secrets: Furniture. 50 real-Life Projects Uncovered. Rockport Publishers, United States of America.
42. Shi, S., and Walker, J. (2006). Wood-based composites: Plywood and veneer-based products. Springer, 391-392. Material: United States Department of Agriculture, United States, Ch. 11, pp. 11.1-11.28.
43. Thomas, E. (1997). Technology scenarios in the Asia-Pacific forestry sector. Asia-Pacific Forestry Sector Outlook Study Working Paper Series, Kuala Lumpur, 80.
44. Thomson, E. (2003). Which hardboard Is best? The Gesso, 7.
45. Ul Haq Bhat, I., Abdullah, C. K., Abdul Khalil, H. P. S., Hakimi Ibrahim, M., and Nurul Fazita, M. R. (2010). Properties enhancement of resin impregnated agro waste: Oil palm trunk lumber. Journal of Reinforced Plastics and Composites 29(22), 3301-3308.
46. Waldemar, J. H., and Jorissen, A. J. M. (2004). Wood modification developments. Heron, 49, 361-386.
47. Youngquist, H. A. (1999). Wood-based composites and panel products. Wood Handbook - Wood as an Engineering Material, U.S, 463.
48. Yusoff, M. N. M., and Koh, M. P. (2001). Production of medium density fibreboard (MDF) from mixtures of oil palm stem and rubberwood. Wood-Based Panel Products, FRIM, Malaysia, Kuala Lumpur, Malaysia, 4.

# CHAPTER 4

# APPLICATIONS OF COMPOSITE MATERIALS IN CONSTRUCTION AND BUILDINGS

## INTRODUCTION AND BACKGROUND

The building or construction industry is considered to be one of the world's biggest consumers of the polymer composites. Construction industries have been utilizing unreinforced polymer materials for a long time in non-load enduring applications, for example, kitchenware, trimmings, cladding and vanities. In the most recent decade, there has been a deliberate thrust to utilize fiber reinforced polymer composites (FRPCs) into the construction industry for use in essential load bearing structures. Prospective benefits of polymeric composite materials include high specific stiffness, extraordinary strength, excellent fatigue performance, tailorable durability, lesser maintenance expenditures and flexible fabrication routes. Considering the unique characteristics of composite materials, FRPCs are being employed in applications such as retrofit and rehabilitation, fiber composite based civil structures and substitute reinforcement for concrete. However, the number of principal structural applications of fiber reinforced polymer composites in construction industry remains comparatively low due to various issues related to construction industries. Issues such as budget, poor knowledge of construction problems by composites industries, the absence of industry standardization and nonexistence of design codes, and deficiency of experienced designers with rich experience in construction and composite industries are generally considered to put these (composite) materials off guard when compared with conventional construction materials. This chapter focusses on advantages of composite materials in construction business over traditional materials.

Composite materials join and retain two or more distinctive phases to deliver a novel material which possesses the characteristics far better than both of the base components. Composites have been utilized in the construction industry for many years. Straw has been used to reinforce and strengthen the brick blocks for more than 2000 years from now, and this technique is still being utilized today. The use of metallic reinforcements in concrete beams has also been evidenced in Greece, almost 1000 years back.

Polymer matrix materials such as vinyl ester, epoxy, and polyester are combined with reinforcing fibers and fillers to form multiphase composite materials with exceptional mechanical and thermal properties. Fillers are added in the composites to enhance the characteristics of composite materials by lowering bulk density, reducing overall cost and improving aesthetic features. Fibrous reinforcements are responsible for improving mechanical properties, such as stiffness and mechanical strength, of the composites. Fiber reinforced composites contain high strength fiber reinforcements of aramid, glass, and carbon which are responsible for withstanding loads, while polymeric matrix binds the fibers and protects them from the environment.

For decades, the construction industry has utilized the polymer matrix composites for nonstructural applications such as vanities, cladding, finishing, and decoration. As the world's second biggest consumer of composite materials, the construction industry represented 35% of the worldwide composite market in 1999 (Weaver, 1999). Recently, the composite materials have gained enormous importance for their use in load-bearing structural applications. Composites materials employed in load bearing civil structures have presented promising and competitive advantages over conventional materials, i.e. already replaced steel structures.

## USES OF FIBER REINFORCED COMPOSITES IN CONSTRUCTION

Currently, composite materials are rarely utilized in critical load-bearing structures but their application in retrofit and rehabilitation of structures very common. Fiber reinforced composites are also employed as a substitute for steel rods in concrete structures. Moreover, most of the civil structures utilize composite components during manufacturing, i.e. composites are an essential part of civil structures.

The deterioration of civil infrastructure in Europe, USA, and Canada is widespread and unwavering (Haile, 2003). The projected expenditure to retrofit and rehabilitate the existing global infrastructure is about $900B (Khabari & Zhao, 2000). Concrete structure in Australia is estimated to be repaired at an annual rate of $500M (ISIS Canada, 1998). External steel sheet or concrete is used as a retrofit to improve the mechanical characteristics (strength, ductility, stiffness) of the existing civil structures. Concrete is a versatile material which has widespread uses in complex on-ground and submerged structures. It is extensively utilized for encapsulation of pier in bridge structures (Oehlers, 2000). Steel bars are combined with concrete mixtures to deliver a composite structure which is used to repair deteriorated concrete infrastructures with a slight addition in weight. The use of composite materials as a rehabilitation and retrofit material has been increasing rapidly. In recent years, composite materials have been used extensively as a protective layer to enhance the mechanical response of civil infrastructures. In most of the retrofit cases, composite fibers such as tow, plates, fabrics, jackets and strips, are typically combined with civil structures externally. Composite materials are flexible in usage as compared to conventional materials because of their ability to be manufactured into complex shapes and bond to various kinds of base structures. Composites also offer an opportunity to tailor the directional properties of materials, i.e. to enhance the strength in one direction while keeping it constant in other directions. However, it is a challenging job to manufacture isotropic composite materials which, like metals, possess similar properties in all directions.

Fiber reinforced plastics (FRP) have been used as reinforcements in concrete in the 1960s. FRP reinforced concrete has been under consistent investigation over the years to improve its utility in civil infrastructures. Multifarious forms of unstressed FRPs have been developed which include carbon fiber bonded with polyester, epoxy bound E-glass, and grooved rod of FRP, E-glass mesh and reinforcing cages made from box sections and flat bars (Carse, 1996). FRPs are also available in the stressed state in the form of FRP strands which run parallel to the tendon axis. FRP strands are utilized in a similar pattern to traditional steel tendons (Gowripalan, 1999). FRP reinforcements have emerged as excellent substitute materials to cater the issue of corrosion in steel reinforcements. Enhanced durability of FRP reinforcements is a deciding factor in their selection as reinforcements in concrete structures (Gowripalan, 2000). FRPs have been reported to have other advantages over conventional reinforcements such as enhanced handling velocity and their suitability for material sensitive applications, i.e. FRPs can be used in radiation environment and electromagnetic applications without disturbing the spectrum of emissions (Ko et al., 1997).

Many load-bearing civil structures have been developed from FRPs over the last few decades. The applications of FRPs include, compound roofs, bridge decks, pedestrian bridges, energy capturing pavement guardrails, building systems, rooftop cooling towers, access platform for chemical and industrial offshores, power poles, light poles, electricity transmission towers and marine erections such as fenders and seawalls (Karbhari et al., 1999; Hollaway, 2002; Bank & Gentry, 2000; Barbero & Gangarao, 1991).

Fiber reinforced composites offer many potential advantages including high specific stiffness and strength, tailorable durability, excellent fatigue resistance, longer life and minimal expenditures. However, the potential benefits of FRPs are, sometimes, not achieved due to the unsuitability of the manufacturing processes and higher equipment costs. Additionally, unavailability of certain composite applications has also made the composite industries to question the capability of composites to replace the conventional materials. Currently, most of the structural applications are using the composites on a trial basis to evaluate the characteristics of composite materials in real-time structural environments. Moreover, some applications require expensive composite materials to fulfill their fundamental needs, rendering the composites as an economically impractical choice.

## Seismic Retrofit of Bridge Columns

The seismic retrofit or rehabilitation of bridge piers is carried out by either winding process or sheet method. Winding process involves the application of unidirectional tape to a concrete surface which has fibers in the longitudinal direction. Wet-winding of fiber tow is carried out circumferentially after the application of tape on a concrete surface. A fire resistant coating material is applied on the structure after circumferential winding and curing of the composite. Fire resistant coating may include calcium silicate, cement mortar or comparable high-temperature polymer material. The purpose of the unidirectional tape is to provide flexural strength while the circumferential fiber tows are responsible for structural confinement. The winding process or sheet method usually deals with applications involving carbon fiber reinforcements. Moreover, this approach has also been utilized to form smokestacks or chimneys. A schematic of the winding process is displayed in Fig. 33. In sheet method, adhesion of tape or sheet is carried out circumferentially or longitudinally to stack the sheets one over another to attain the desired thickness. The entire sheet process is similar to wet layup process which involves application and impregnation of resin manually using hands, squeegees or rollers. This method is commonly utilized for application tow sheet, Replark and fiber cloth (Karbhari, 1998).

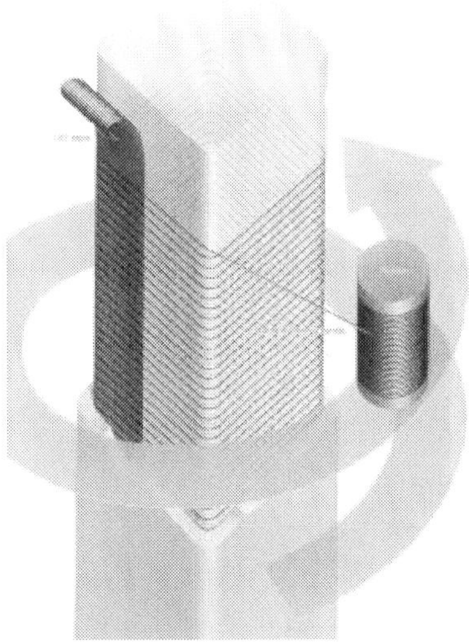

**Figure 33:** Schematic of the "winding" method for column/pier retrofit

The retrofit or rehabilitation philosophy vary in different regions of the world, including Japan and USA. Japanese retrofit philosophy focusses on providing the strength and stiffness to the column instead of improving deformation capability or ductility. Additionally, Japan has a substantial number of single column bents with longitudinal reinforcements, which are detached along the height because of the cantilever reaction without taking into consideration the tension stiffening in cracked zones. Vertically aligned fibers are primarily employed to address the issues related to column bents. Aramid based fiber systems utilize both fibers and fabrics in retrofit applications. In aramid based systems, the application of sheets is carried out by manual layup process; similar to carbon fiber sheet applications.

Replark sheets are extensively utilized for retrofit of columns fractured by seismic/catastrophic failures. Two layers of Replark sheet are applied for shear strengthening in a vertical direction and flexural reinforcement in a circumferential direction. Vertical alignment of Replark sheets is necessary to provide strength to existing columns for facilitation of higher traffic flow. FORCA tow sheet was used for retrofit of a cracked column on the Yodogawa Bridge Japan. The retrofitted area is shown in Fig. 34. The retrofit was conducted by placing a carbon tow sheet circumferentially as a band instead of applying it over the total height. The pier/column of the bridge, having a variable cross-section, is supported by four individual pillars which are joined by diaphragm walls. The foundations of the pier placed in water.

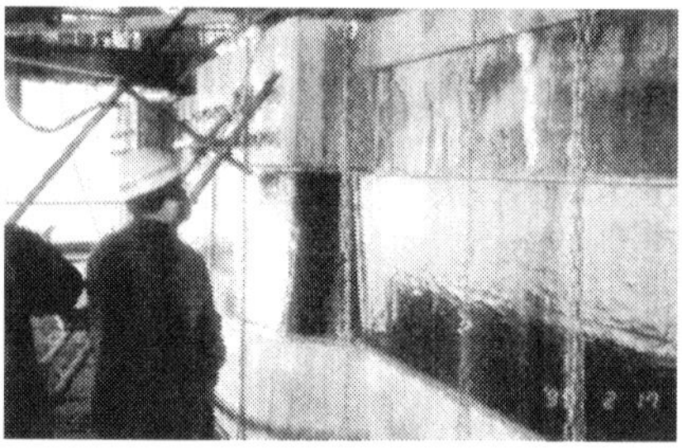

**Figure 34:** View of band-retrofitted pier of Yodogawa Bridge

In 1996, the use of Tonen's tow for retrofit applications was about 55% of the total sale (Tonen Corporation is a composite manufacturing company in Japan) while the use of Mitsubishi Chemical's product called Replark was recorded to be 12.5% of the total sale.

## External Strengthening of Deck Soffits and Bridge Girders

In recent times, carbon fiber sheets have been used expansively for the repair of deck softies and bridge girders in Japan and USA. The use carbon fiber sheets for bridge strengthening was initially carried out in 1984 by Tonen Corporation. The use of composite materials in bridge strengthening applications depends on the following criteria:

- Strengthening of the bridge to enable greater wheel loads without hitting shear
- Enhancement of the existing longitudinal deck supports

Carbon fiber sheets are used to arrest further cracking of decks, enhance the load carrying capability and strengthen the sound walls. Fig. 35 shows the application of carbon fiber tows for repair of a cracked deck soffit.

**Figure 35:** Application of longitudinal tow sheet to repair a fractured deck soffit

There are two methods for application of the sheets in repairing of bridge structures. The first method involves the continuous use of sheets without any uncovered area of concrete. This method provides a uniform coverage of the sheet material on the deck structure as shown in Fig. 36.

**Figure 36:** Application of continuous Replark sheet for the strengthening a road deck

The second method for bridge repairing applications is called Sho-Bond method which was invented by Sho-Bond Corporation Japan. This process involves the bonding of CFRP layers to a concrete structure to form a lattice design, as shown in Fig. 37. This structure allows the observation of a crack initiation and facilitates the escape of water which permeates into the concrete structure. Water needs to flee from the concrete composite structure to save the bridge from degradation and peel. The Sho-Bond method utilizes two distinct constituents of resin systems which have different colors. The mixing of dark pink resin with a hardener results in a pale yellow mixture. The change of color is an indication of proper mixing of resin systems.

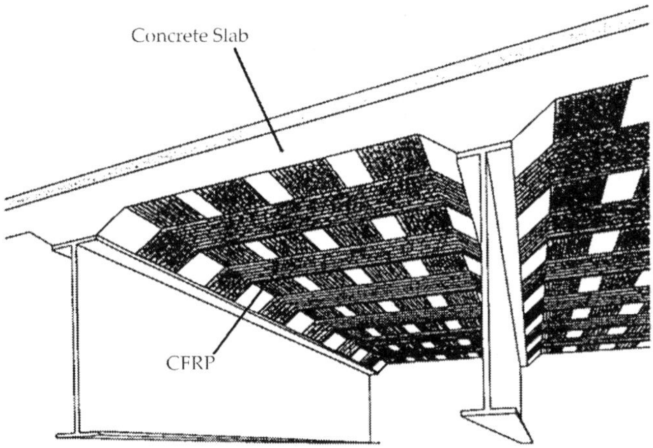


**Figure 37:** Application of the lattice pattern formed by Sho-Bond method

Irrespective of the repair process, the concrete surface must be cleaned, filled with a paste or mortar to reconstruct a uniform surface, and concealed with a primer before the application of fibers by adhesion or layup process. In most of the case, epoxy putty or mortar is applied on the primed surface to smoothen the surface. Most of the times, the curing of composites is carried out in ambient environmental conditions without the incorporation of vacuum environments for achieving compact structures. The Sho-Bond process was initially designed for Torayca cloth systems, but this method actively supports the use of Tonen's tow sheet for bridge repairing applications. Modified Sho-Bond method (with undefined lattice pattern) was employed for retrofit of Hinobashi Bridge section in 1993, as shown in Fig. 38. The bridge experienced cracks in the region between steel girders. The direction of the cracks was perpendicular to the supporting girders. Single tow sheet of carbon fiber (FORCA FTS-C1-20) was applied in the longitudinal direction of the deck. Application of two sheets was carried out at four areas with an overlap of 10 cm between adjoining fabric layers. Load measurement tests were carried out before and after the rehabilitation/retrofit to evaluate the load bearing capacity of steel bars present in the structure. It was recorded that the incorporation of tow sheets reduced the stress concentration of the steel bars by 35 %. The Sho-Bond method did not interrupt the flow of traffic during the retrofit process.

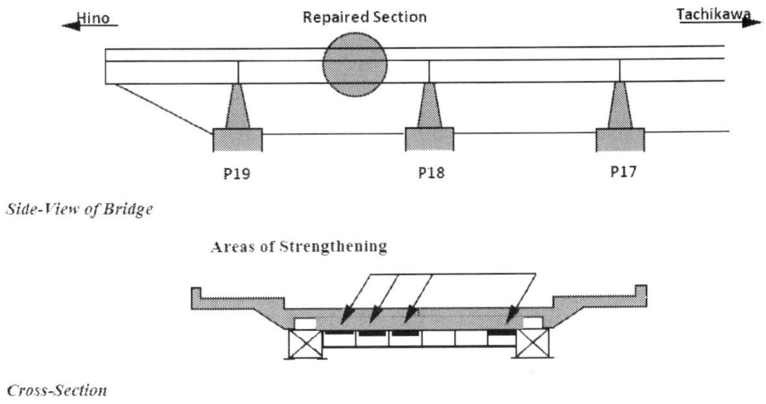


**Figure 38:** Schematic of the modified Sho-Bond retrofit at the Hinobashi Bridge.

The utility of tow sheets in the repair of bridge girders accounted for 15% of the Tonen's sheet material sale in 1996. Similarly, the use of Mitsubishi Chemical product, Replark in the repair of bridge girders was around 10% of the total Mitsubishi Chemical's sales between 1987 and 1995. Toray's cloth was also used generously (>8,000 m$^2$) for deck repair applications in 1996. Most of the composite material applications were meant for crack control, enhancement of deck capacity for new sound barriers on cantilevers, preclusion of exfoliation in concrete and increment of load levels (Karbhari, 1998).

## Retrofit of Buildings

Unidirectional sheet materials are an attractive choice for retrofit of building structures due to their adaptability and flexibility to be utilized in smaller areas and closed spaces without a requirement of significant structural changes in buildings. The minimum disturbance is faced by the residents of the buildings during the retrofit operations. Additionally, the composite materials for retrofit applications are typically maintenance free due to their unique properties such as corrosion resistance and inert to chemicals. Composite retrofits can be concealed with decorative coatings to improve the aesthetic features of the building structures.

Mitsubishi Chemical's Replark fiber sheet has been used widely in retrofitting of building structures. Between 1987 and 1995, almost 59.3% of the total Replark consumption was used for building retrofits. About 12% of the Tonen's tow sheet (FORCA) was utilized for the retrofit of buildings in 1955. Today's Toraya cloth was also reportedly used in substantially higher amounts for building retrofits. Aramid fiber has also been used in limited amounts for retrofit applications. These sheet materials have been utilized for a range of structural components including seismic retrofit of building columns, repair of cracked floorboards, strengthening and reinforcement of heavily deteriorated columns in commercial buildings, seismic reinforcing of walls and columns in pinball parlors, support of construction and parking structures by slab and external strengthening of columns in buildings. In most of the cases, the sheet materials have been applied for retrofits of buildings as preventive measures after the observation of initial cracking. However, retrofits are also used for strengthening deteriorated building structures. The application of Replark sheet for retrofit of columns in a commercial office building is shown in Fig. 39.

*(a) Before retrofit* *(b) After wrapping* *(c) After finishing*

**Figure 39:** Retrofit of deteriorated columns by Replark sheets

## Strengthening of Tunnels

Tunnels are used extensively in mountainous and uneven topographic regions, for highway and railway systems. For instance, Japan has around 4,600 tunnels for railway tracks with a total tunnel length of approximately 2,900 kilometers. Around 6,500 road tunnels in Japan are collectively 18,000 kilometers long. Additionally, tunnels are utilized expansively in commercial hydroelectric projects for supply of water and for providing access to underground water storage facilities. A huge number of underground tunnels were made before and shortly after WWII. Old tunnels have now completed their design lifecycles, especially their linings and supports.

Currently, the tunnel strata have become weaker and deteriorated due to some flaws resulting in severe deformation and stresses on the tunnel linings. Around 45% of the tunnels require repairing to fend off the risks of tunnel failures. Principle types of tunnel defects include deterioration of tunnel linings, inadequate headspace due to distortion, dislocation of linings, and excessive seepage of ground water into the tunnels. Traditional retrofits have been used extensively by using rock bolts, steel framings, and concrete. However, railway tunnels and highways require quick retrofit methods because of the availability of shorter time spans between traffic flows. Composite materials are an ideal choice to deliver the useful retrofit applications for railway and highway tunnels.

Considering the application for the strengthening of tunnel lining, carbon fiber sheet materials have been utilized for retrofit of tunnel linings and interior surfaces, similar to their use in column applications. Retrofit of carbon fiber sheet includes the preparation and priming of the surface, coating of epoxy on the primed surface and wrapping of epoxy impregnated. Retrofit of tunnel lining is shown in Fig. 40.

**Figure 40:** Application of epoxy resin on FORCA fiber sheet for retrofit of tunnel lining

The retrofit method discussed above is usually given precedence over other techniques due to its convenience, fast processing and capability to carry out retrofit with minimum traffic disruptions. Fig. 8 exhibits the use of carbon fiber sheet (FORCA FTS C-200) to repair the concrete linings in a tunnel. The application of layers in the retrofit of tunnel linings was carried out longitudinally. In the current case, the retrofit was performed by carbon fiber sheets after the tunnel had gone through water seepage and consistent static loads.

Retrofit of the inner lining of Nikkureyama Tunnel (located in Japan) was carried out by the placement of carbon fiber tow sheet (FTS-C1-20 FORCA). The details of the Nikkureyama Tunnel retrofit are shown in Fig. 41. The retrofit operations were carried out by Tonen Corporation and Sanshin Construction Co. Ltd. in 1992. The purpose of the retrofit was to repair the cracked section of the concrete arch and to preclude further exfoliation and deterioration of concrete due to an upsurge in rock pressure arising from an extensive mudstone layer. The retrofit was also performed to reinforce the tunnel lining to sustain additional rock-pressures over the next two decades. The area of the tunnel retrofit was measured, and it was about 40 meters in length and 9 meters across the arch. Before the application of single tow sheet layer, the existing lining surface was prepared by grinding the surface (for maximum adhesion) and filling the voids with resin injection. The arch had undergone substantial deformation before the application of retrofit. However, the arch had reached the equilibrium state where further deformation was not expected in near future Karbhari, 1998).

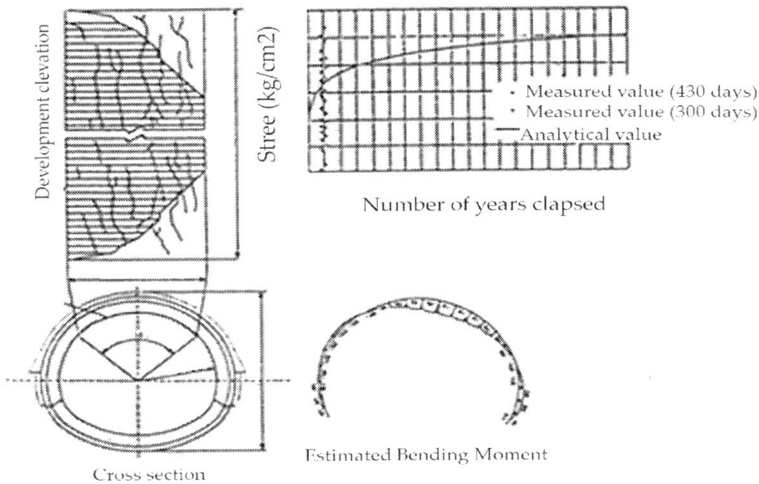


**Figure 41:** Schematic of performance estimates and cross-section of the Nikkureyama Tunnel.

Retrofit of Ningyotoge Tunnel (Japan) was carried out by Nikohu Construction Co. in 1993. The sidewall in the tunnel was deteriorated due to its continuous contact with metaphoric rock, which eventually, resulted in the weakening of arch section. Initially, a primer layer was deposited on the arch surface after grinding and resin filling of cracks and irregularities. The A primer capable of being applied to a wet surface was used on the substrate after grinding off irregularities and filling crevices and cracks with mortar. Fig. 42 shows the application of two composite layers on the arch section of the tunnel. However, a third layer was also applied to localized regions, which exhibited severe cracking. A fluorinated paint layer was applied on the composite layers (consisting of FR-E3 resin and FTS-C1-20 carbon sheets) after their curing. Similar to the retrofit of Nikkureyama Tunnel in Japan, the main purpose of the Ningyotoge Tunnel retrofit was to improve the load bearing capacity of the tunnel and to preclude further exfoliation and cracking of concrete.

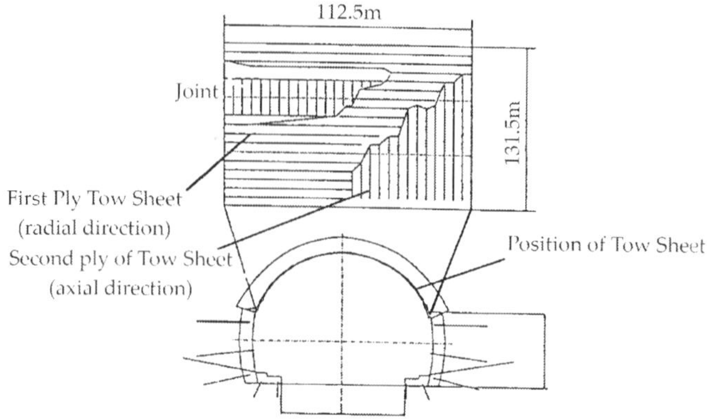


**Figure 42:** Schematic of the Ningyotoge Tunnel retrofit application

Sometimes, the retrofit of the tunnels is performed to enhance its load bearing capacity due to the construction of road above it. Due to high stiffness of carbon fibers, most of the tunnel retrofit applications utilize carbon fiber sheet material (FORCA), fabricated by Tonen Corporation. It was reported that almost 11% tow sheets of Tonen Corp. were used for retrofit applications of tunnels in 1996.

## Retrofit of Chimneys and Smokestacks

Obayashi Corp. and Mitsubishi Kasei pioneered the utilization of carbon fiber sheet materials for retrofit of chimneys and smokestacks in 1987. Wet winding process for composite materials gained vital importance for retrofit applications of smokestacks and chimneys in the late 1980s. The wet winding process involves the longitudinal use of carbon fiber prepregs in the structure followed by mechanized wet-winding of fiber tows (carbon fiber) in the hoop direction. The prepreg-fiber layer is allowed to cure after its application onto the chimney walls, and fire resistant layer is coated on the composite coating. Reports have unveiled the inconsistencies in an automated wet-winding process due to the incorporation of ring setting. Therefore, the wet winding process has, now, been replaced by carbon fiber sheet composites which are applied by wet-layup technique as used in column retrofit applications. Tonen's FORCA tow sheet and Mitsubishi's Replark sheets are most suitable materials for retrofit applications of civil infrastructures. Application of Replark sheet for retrofit of Chimney is shown in Fig. 43.

**Figure 43:** Application of Replark sheet onto a chimney

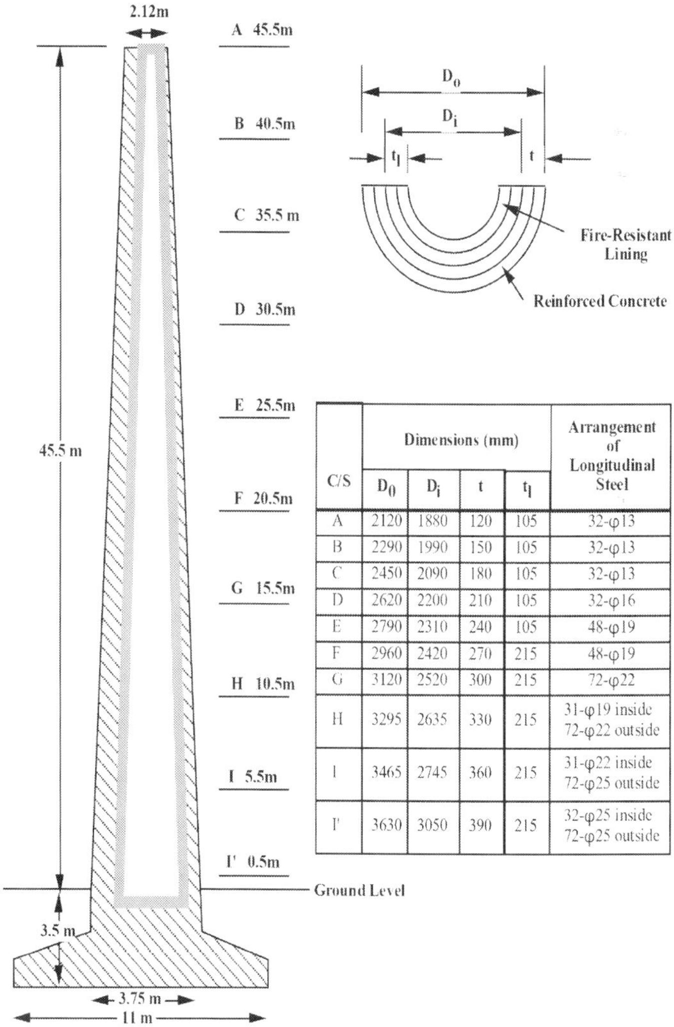


| C/S | Dimensions (mm) | | | | Arrangement of Longitudinal Steel |
|---|---|---|---|---|---|
| | $D_0$ | $D_i$ | t | $t_l$ | |
| A | 2120 | 1880 | 120 | 105 | 32-φ13 |
| B | 2290 | 1990 | 150 | 105 | 32-φ13 |
| C | 2450 | 2090 | 180 | 105 | 32-φ13 |
| D | 2620 | 2200 | 210 | 105 | 32-φ16 |
| E | 2790 | 2310 | 240 | 105 | 48-φ19 |
| F | 2960 | 2420 | 270 | 215 | 48-φ19 |
| G | 3120 | 2520 | 300 | 215 | 72-φ22 |
| H | 3295 | 2635 | 330 | 215 | 31-φ19 inside<br>72-φ22 outside |
| I | 3465 | 2745 | 360 | 215 | 31-φ22 inside<br>72-φ25 outside |
| I' | 3630 | 3050 | 390 | 215 | 32-φ25 inside<br>72-φ25 outside |

**Figure 44:** Dimensional details of Wakayama structure

In 1991, Wakayama chimney project utilized a double layer of carbon fiber sheets, which were applied in both longitudinal and circumferential direction. Severe cracking of concrete structure and excessive corrosion of steel reinforcements were driving factors for retrofit of the chimney. Formerly, steel bands were installed in lateral and longitudinal direction for arresting the cracks (0.3 mm to 10 mm in length) and strengthening the structure. Surprisingly, the installed steel bands were unable to fulfill the requirements (control corrosion and arrest cracks) and resulted in deterioration of the structure due to enhanced cracking of concrete and corrosion of steel. The chimney was around 45.5 m high with outer diameters of 2.12 m and 3.63 m at the top and bottom respectively. The dimensions of the chimney are shown in Fig. 44.

Initially, the steel bands were removed, and the cracks were filled with resin before the application of fiber tows onto the concrete surface. The design of the retrofit was made to provide additional strength (more than that provided by steel) to the structure. Comparison of strengths for individual constituents and the composite material is shown in Table. 4. It is clearly visible that fiber reinforcements show exceptional strength as compared to steel reinforcements Karbhari, 1998).

**Table 4:** Characteristics of retrofit materials

| Material | Performance Criteria | Specified Strength ($kg/cm^2$) | Short-Term Tolerable Stress ($kg/cm^2$) | Elastic Modulus ($kg/cm^2$) |
|---|---|---|---|---|
| Carbon Fiber Tow Sheet (t = 0.2 mm) | Tension | 14,000 | 9350 | $1.4 \times 10^6$ |
| Steel | Tension & Compression | 2400 (yield) | 2400 | $2.1 \times 10^6$ |
| Concrete | Compression | 150 | 100 | $2 \times 10^5$ |

The stress levels experienced by composite reinforced structures are usually low due to their ability to withstand high stresses without deformation. The retrofit of Wakayama chimney was carried out by Hanshin Construction Co. and Taisei Construction Co.

It was reported that almost 28 sites were retrofitted by using carbon fiber sheets and Mitsubishi's Replark sheets from 1987 to 1994 in Japan. Moreover, almost 2% Tonen's tow sheets were utilized for retrofit applications of chimneys in 1996. Mitsubishi spent around 18% of its products for chimney retrofit applications between 1987 and 1995. In addition to the utilization of carbon fiber tow sheets, aramid fiber sheets have also been used for retrofit of chimneys. Both automated and hand layup methods were used for impregnation of fiber tows in the resin systems. However, a combination of automation and manual skills is required for achieving desired structural properties in retrofits.

## Retrofit of Other Structures

Primary applications of fiber tows and sheets include bridges, buildings, chimneys, and tunnels. However, composite materials can also be utilized for repair and seismic strengthening of various structures. Formerly, longstanding historic structures were retrofitted by conventional reinforcing materials (concrete and steel), but the use of modern composite materials have revolutionized the idea of retrofit due to its adaptability, convenience and high aesthetic value. Flexible sheets of composite materials are extremely suitable to use for retrofit applications because of their less susceptibility to cracking. Some of the typical applications of composite sheets in civil sector are given below:

- Modern composite layers are used to strengthen, repair, and upgrade stiff retaining walls of concrete, concrete foundations and pedestals, reinforced sewer pipes, concrete water tanks and pre-stressed pipes of concrete materials. Composite structures have been used in strengthening, waterproofing, and repair of box culverts and water channels.
- Composite sheets are used abundantly for applications in wooden electric poles and side-rails of walkways.
- The building is made shock proof by incorporating a blend of carbon fibers and glass fibers in the structure.
- Owing to its corrosion resistance and ability to be installed rapidly, composite sheets are being used extensively for retrofit of marine structures, such as quays and dock walls.
- Glass fiber sheets are excellent candidates for repair of block

balustrades walls. A single fiber sheet is enough to strengthen the corroded and deteriorated sections (Fig. 45).

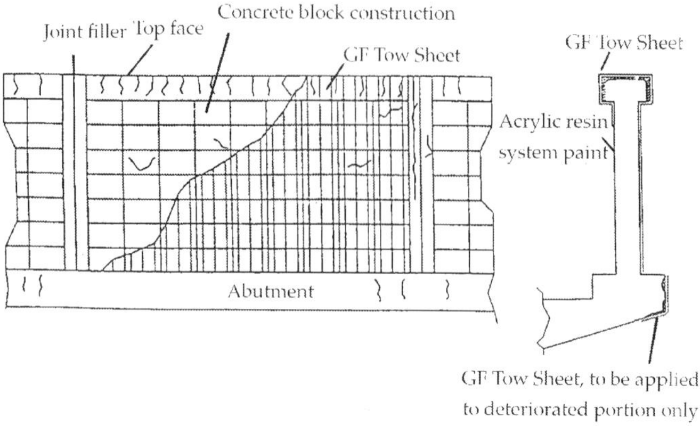


**Figure 45:** Schematic of repair of a concrete structure using FORCA glass sheet.

## ISSUES ASSOCIATED WITH THE USE OF COMPOSITE MATERIALS IN CONSTRUCTION INDUSTRY

Despite the exceptional structural properties of composite materials, their application in the construction industry is growing at a slow rate. There are various reasons for slow ingression of composite materials in the construction sector. Some issues of the composite materials related to their cost, durability and structural performance are illustrated below.

### Cost

Numerous expenditures are associated with the use of composite materials in the civil sector. Two kinds of costs are incurred for application of composites in the construction industry, i.e. short term cost and long term cost. Short-term costs include the expenditures incurred during design, installation and construction phases while long term costs are incurred for disposal, maintenance, modification and deconstruction of structures. Composite expenditures can be further categorized into a direct cost (material and production costs) and indirect cost (cost incurred in depreciation, environmental impacts, and interruptions in traffic and resale value). Determination of the minimum value for a particular application is a challenging job as the cost estimate may vary based on the quality of the products.

## *Short Term Cost of Fiber Reinforced Composites*

Presently, fiber reinforced composites are comparatively costlier than the conventional structural materials on the basis of initial costs. Some factors are responsible for controlling the cost of composites such as expensive raw materials, costly equipment, low availability of high strength aerospace materials and the use of novel imported materials (Hale & Gibson, 1997).

Composite materials are being extensively used in sports industry due to their light weights with improved stiffness, and their cost is declining rapidly. Conversely, civil and structural applications of composite materials are quite costly, but researchers believe that their value will also fall after the increment in their production volumes (Goldstien, 1996). However, the cost of composite materials also depends on the origin of composite materials, e.g. Australia relies on the imported composite materials for construction applications. The import of goods is subjected to some cost variables which affect the overall cost of the structural composites, i.e. Fluctuations in international manufacturing costs, import and transport costs, and variations in the exchange rates between different countries. All above mentioned factors in combination with the reduction of cost based on the quantity of the purchased are responsible for controlling the final composite costs. However, it is quite hard to predict the exact costs for any composite product because of the involvement of multiple controlling factors.

## *Manufacturing cost*

In addition to comparatively higher material costs, the short-term expenditures of FRPs also depend on the manufacturing processes. Initially, most of the composite manufacturing processes were developed for aerospace, automotive and marine industries. However, the construction sector has different parameters for composite applications as opposed to other industries. Construction industries deal with bulk material structures which are designed to sustain extreme loading and environmental conditions. Additionally, the design specifications of civil structures cannot be completely duplicated due to novel designs for different applications. Therefore, most of the construction projects are carried as 'one-off' tasks. The construction industry is different than the production industries which involve mass production of specific designs parts. Consequently, the common design fabrication processes in production industry are not suitable for the construction industry and civil sector.

Mass production of some light weight civil components may result in the reduction of short-term costs. Lightweight structural components reduce the carnage and transport costs while bulky modular components of composites are excellent candidates to reduce the installation time. Meier (2000) discussed the impact of indirect cost savings on the overall cost of composite materials. He was of the opinion that the application of composite materials certainly reduces the production and installation cost. However, the exact cost cannot be quantified. Due to the usage of lightweight material structures and flexible equipment, the faster construction of the structures is possible which results in lesser distress and disturbance in the environment. Ehlen (1999) and Shapira et al. (1997) claimed that composite materials could reduce the production and maintenance cost of composites used in road closures, civil structures, sheds of parking lots.

## *Cost Estimation of Fiber Reinforced Composite Materials*

Techniques used for cost analysis of civil infrastructures may vary depending on the nature of the projects. Most of the times, the cost analysis of the building projects is based on the initial costs of raw materials and processing equipment. This technique is mainly based on the fact that the owner requires the best possible building structure at a minimum possible cost, irrespective of the long-term performance of the structure.

There are two major techniques which address the critical issues associated with the application of composite materials in civil structures. The first technique was developed by El-Mikawi and Mosallam (1996) which utilizes the Analytical Hierarchy Process (AHP) to evaluate choices for critical requirements, manufacturing, management and maintenance of the projects. However, AHP technique does not provide the comparison of tangible and minor factors, including influence on amenity, or cost analysis of the available options for evaluating the realistic composite materials. The second technique for cost estimation is called Whole of

Life (WOL) method which is derived from lifecycle cost analysis (Ehlen, 1999). WOL is a comprehensive analysis technique involving initial cost, refurbishment, and replacement costs, operating cost, maintenance cost, disposal costs and other miscellaneous costs such as depreciation, taxes, and management costs. A particular value is assigned to the ongoing cost of building structures based on the projected lifetime of the buildings. WOL is a useful tool which can be employed to compare the benefits of existing and new materials in building structures. The comparison of the materials is based on the similar value of critical performance factors such as strength, stiffness and minimum service life.

## Structural performance

### *Specific stiffness and specific strength*

Fiber reinforced polymer composites have different applications in the construction industry due to their high specific stiffness and specific strength. FRP composites provide superior mechanical properties along with considerable reduction in weight. The major benefits of low weight composite materials include the liberty to manufacture larger components, reduced size of foundations and substructures, and reduced transportation costs.

Careful utilization of composite materials as a substitute for conventional structural materials can result in the production of structures with high strength to weight ratio. Lighter materials are cheaper than the conventional materials. However, this kind of multidimensional materials is slightly difficult to produce. Presently, the manufacturing of a composite bridge deck may cost ten times the cost of conventional concrete materials. The utilization of lightweight composites in deck applications can reduce the transportation cost up to 75 %. However, the cost saving for lightweight components is not very high if the installation site is located adjacent to the manufacturing site. Reduction in lifting cost can also contribute towards cost saving for lightweight composite materials.

### *Low stiffness*

Most of the civil structures require high strength and stiffness for load bearing applications. However, high stiffness FRP composites may become substantially overdesigned rendering them economically less competitive. Moreover, high stiffness carbon fiber reinforced polymer composites are comparatively expensive, and their utilization is limited because of lower failure strains. It is estimated that the cost of the high strength and high stiffness composite laminates will decrease significantly owing to their bulk production with increased fiber volume fractions.

### *Tailorable mechanical properties*

Fiber reinforced composite materials are considered to be versatile as compared to the conventional structural materials due to their flexibility to be engineered in various ways. The composite materials with versatile characteristics can be manufactured by maneuvering the fiber and resin fractions in the mixture, to achieve suitable performance parameters such as stiffness, strength, durability, fatigue resistance, impact strength, and eco-friendly behavior.

A flexible fabrication process is required to manufacture a composite structure with all the desired characteristics. Localized variation in fiber and resin content can also be achieved by utilizing the suitable manufacturing process. Automated manufacturing processes such as Pultrusion usually permit the variations in reinforcing fiber types, however localized changes in resin application is not supported in this process. Modifications in the fabrication processes are required to manufacture the composite materials with tailored properties.

### *Durability*

The major advantage of composite materials over traditional materials is their excellent durability in under real-time environmental conditions (Hayes et al., 2000). There is a wide range of materials which are suitable for manufacturing of structures with superior properties such as excellent resistance to long-term deficits which may result from moisture, freeze-thaw cycles, chemical attack, ultraviolet radiation, dynamic loading, and deterioration properties by aging.

FRP composites have been under service conditions, in civil and construction applications, for a very short duration of time. Therefore, it is difficult to accurately assess their durability. It has been observed that some FRP composites have shown affordable durability in infrastructure and civil applications (Liao et al., 1998). However, the long-term durability data (service life of 75-100 years) for composite based civil structures is not available (Karbhari & Zhang, 1999). An international research was conducted by Civil Engineering Research Fund (CERF) for bridging long-term and short-term durability data. The results of the study are based on the evaluation of key composite performance parameters such as alkaline solution, moisture effects, creep, fatigue, and physical degradation of materials (Karbhari, 2001).

## REFERENCES

1. Bank, L. C., & Gentry, T. R. (2000). Composite materials for roadside safety structures. In Proceedings of ACUN-II International Composites Conference, University of New South Wales, Australia (pp. 68-77).
2. Barbero, E., & Gangarao, V. S. (1991). Structural applications of composites in infrastructure.
3. Carse, A. (1996). The asset management of a long bridge structure affected by alkali-silica reaction. In International Conference on Alkali-Aggregate Reaction in Concrete, 10th, 1996, Melbourne, Victoria, Australia.
4. Ehlen, M. A. (1999). Life-cycle costs of fiber-reinforced-polymer bridge decks. Journal of Materials in Civil Engineering, 11(3), 224-230.
5. Ei-Mikawi, M., & Mosallam, A. S. (1996). A methodology for evaluation of the use of advanced composites in structural civil engineering applications. Composites Part B: Engineering, 27(3-4), 203-215.
6. Goldstien, H. (1996). Catching up with Composites, Civil Engineer, March, p47, ASCE, New York
7. Gowripalan, N. (1999). Fibre reinforced polymer (FRP) applications for prestressed concrete bridges. In Proceedings of ACUN 1 Conference Composites: Innovations and Structural Applications, UNSW, Australia.
8. Gowripalan, N. (2000). Design Considerations for Prestressed

Concrete Beams with Fibre Reinforced Polymer (FRP) Tendons. In AUSTROADS BRIDGE CONFERENCE, 4TH, 2000, ADELAIDE, SOUTH AUSTRALIA (Vol. 3, No. AP-G64/00).

9. Haile, S. M. (2003). Materials for fuel cells. Materials Today, 6(3), 24-29.
10. Hale, J. M., & Gibson, G. A. (1997). Strength reduction of GRP composites exposed to high temperature marine environments. In Proceedings of ICCM11 International Conference on Composite Materials, Australian Composites Structures Society, Australia.
11. Hastak, M., & Halpin, D. W. (2000). Assessment of life-cycle benefit-cost of composites in construction. Journal of Composites for Construction, 4(3), 103-111.
12. Hayes, M. D., Lesko, J. J., Haramis, J., Cousins, T. E., Gomez, J., & Masarelli, P. (2000). Laboratory and field testing of composite bridge superstructure. Journal of Composites for Construction, 4(3), 120-128.
13. Hollaway, L. C. (2002). 1.1 The development and the future of advanced polymer composites in the civil infrastructure. In Advanced polymer composites for structural applications in construction: Proceedings of the first international conference, held at Southampton University, UK, on 15–17 April 2002 (pp. 1-21). Thomas Telford Publishing.
14. ISIS Canada. (1998). ISIS Annual Report 1997/1998, The University of Manitoba, Canada
15. Karbhari, V. M. (1998). Use of composite materials in civil infrastructure in Japan. WTEC report. International Technology Research Institute, World Technology (WTEC) Division.
16. Karbhari, V. M. et al. (1999). Structural Characterization of Fiber-Reinforced Composite Short- and Medium-Span Bridge Systems, Developments In Design Standards, CRC-ACS, Australia.
17. Khabari, V. M., & Zhao, L. (2000). The Use of Composite for Twenty-First Century Civil Infrastructure. Fibre Composites: Who Needs Them?
18. Ko, F. K., Somboonsong, W., & Harris, H. G. (1997). Fiber architecture based design of ductile composite rebars for concrete structures. In Proceedings of the 11th International Conference of

Composite Materials (Vol. 6, pp. 723-730).

19. Karbhari V.M. and Zhang S. (1999). Durability of Fibre Reinforced Composites in Civil Infrastructure – Issues, Results and Implications, Developments in Design Standards for Advanced Composites in Infrastructure Applications, CRC-ACS, Australia.
20. Karbhari, V. M. (2001). DURABILITY OF FRP COMPOSITES IN CIVIL INFRASTRUCTURE--MYTH OR REALITY. In FRP Composites in Civil Engineering. Proceedings of the International Conferences on FRP Composite in Civil Engineering. (No. Volume 2).
21. Liao, K., Schultheisz, C. R., Hunston, D. L., & Brinson, L. C. (1998). Long-term durability of fiber-reinforced polymer-matrix composite materials for infrastructure applications: a review. Journal of advanced materials, 30(4), 3-40.
22. Meier, U. (2000). State-of-the-Practice of advanced composite materials in structural engineering in Europe with emphasis on transportation. In INTERNATIONAL COMPOSITES CONFERENCE (ACUN-2), 2ND, 2000, SYDNEY, NEW SOUTH WALES, AUSTRALIA.
23. Oehlers, D. J. (2000). Development of design rules for retrofitting by adhesive bonding or bolting either FRP or steel plates to RC beams in bridges and buildings. In INTERNATIONAL COMPOSITES CONFERENCE (ACUN-2), 2ND, 2000, SYDNEY, NEW SOUTH WALES, AUSTRALIA.
24. Shapira, A., & Bank, L. C. (1997). Constructability and economics of FRP reinforcement cages for concrete beams. Journal of Composites for Construction, 1(3), 82-89.
25. Weaver, A. (1999). Composites: world markets and opportunities. Materials Today, 2(1), 3-6.

# CHAPTER 5

# USES OF COMPOSITE MATERIALS IN ENERGY SECTOR BACKGROUND

It has been reported that international market of composite materials reached 56 billion dollars in 2007. According to the data shared by American Composites Manufacturing Association (ACMA), the United States has around 30% of the total market share. Almost 125,000 people are currently employed alone by the U.S composite manufacturers. Total weight of composite materials manufactured in the U.S exceeds 4 billion pounds. Most of the compounds (around 45% of total) were manufactured and used by the construction industry. 20% of total compounds were used in marine applications, 12% for corrosion, another 12% in the transport industry, 5 % in sports products and 2% were employed by aerospace industry (Greene, 2008). Although the percentage is quite lower for sports products, and aircraft industry, but, in terms of cost, it presents considerably higher percentage as compared to the products used by other industries. The reason for this difference is the complexity of both material and product used in these industries. Unlike other materials which show fluctuation in their market segment, composites are constantly expanding, having more expansion than aluminum or steel products. However, due to the prevalent economic situation, the construction and marine industry are facing downfall regarding composite material manufacturing. The U.S has a distinction of manufacturing several largest composite structures in the world. Few examples include surface sonar domes that weigh more than 20,000 lbs, and a facility located in Jacksonville has manufactured composites heavier than 100,000 lbs. (Greene, 2008). Plans of building 186-foot long motor yacht are in circulation. This large yacht will be built in a new facility located in Tennessee. Seeing this progress of the U.S companies in the field of composites, a plan of manufacturing large structures for renewable energy products is worth giving attention to.

**Figure 46:** U.S. large composite hull fabrication

Production of wind turbine blades from composites is an interesting idea as it is quite possible to produce complex and light weight airfoils from composite materials with least effort. In reality, manufacturing technology and engineering behind this is quite advance and leaves room for the scope of improvement in design and manufacturing processes of larger blades.

There has been growing advancement in the field of renewable energy. The light weightiness of composites makes them a good substitute to their metallic counterparts in large renewable energy devices since the cost of transport and assembly of the devices is greatly reduced using composite materials. Cost is an important parameter especially for large-scale energy projects and composites offers a much reasonable and cost effective solution. We, therefore, have an emerging market that will require huge structures capable of surviving environmental degradation. An important composite compound is Polymer Nanocomposites which displays exclusive physicochemical properties which are unattainable in the case of an individual component. Researchers are especially interested in exploring versatile applications of Polymer Nanocomposites in the fields of storage, EM absorption, automotive and information industry, in military defense, to devise catalysts and much more. We can say that there is a vast scope of research in applications of Polymer Nanocomposites to solve problems faced by energy and environmental sector.

This chapter mainly presents an overview of the latest advances in polymer Nanocomposites to be used in saving energy (electrochemical devices), anticorrosion (anticorrosive coatings), and energy storage (capacitors and batteries) applications.

## USAGE OF COMPOSITES IN ENERGY SECTOR

Renewable energy devices can be divided into two main parts:

- Large public projects which provide major share of our energy
- Energy devices at household level

The main advantage of renewable energy is that its generation doesn't require any central production plant. Composites have applications at both industrial and household level.

Composites allow the production of light weight and repair free modules to be used in large structures. Corrosion resistance is of utmost importance for ocean renewable projects. Mass production of complex structures from molds of different kinds in a cost effective manner is indeed a smart idea. Designers are now fully aware that complicated composite geometric structures can be easily reproduced. This trait of composites is thus effectively aiding scientists to think outside the box and produce nonplanar modules having lesser assembled parts.

### Wind Turbine Blades

Installment of the wind turbine is increasing worldwide; in the current year, around 15,000 more turbines were installed than the previous year. To minimize transportation costs, manufacturers like to produce blades as close to wind farm sites as possible. Blades of these turbines were made of composite materials which give a total weight estimate of around 44 million pounds composites used for this process (Wiser & Bollinger et al., 2006). In 2007, the share of the U.S in this market was reported to be around 27 % (Laxon & Hand et al., 2006).

The following table shows data for some manufacturing facilities recently developed in the U.S. these facilities have generated nearly 1 billion dollars of international investment.

**Table 5:** Recently Developed Wind Turbine Blade Production Facilities in the U.S.

| Manufacturer | Plant Location | # of Employees | Investment | Internet Reference Source |
|---|---|---|---|---|
| Acciona Windpower | West Branch, Iowa | 100 | $23,000,000 | http://www.westbranchtimes.com/article.php?id=1689 |
| Fuhrländer AG | Butte, Montana | 150 (housings)<br>600 (blades) | $25,000,000 | http://www.fuhrlaender.de/index.php?option=com_content&task=view&id=94&Itemid=104&lang=en |
| Gamesa SA (Fiberblade LLC) | Ebensburg, Pennsylvania | 235 | $50,000,000 | http://www.gamesa.es/en/press/press-releases/gamesa-opens-its-first-high-tech-manufacturing-plant-with-an-investment-of-50-million-dollars |
| Knight & Carver | Howard, South Dakota | 50 | | http://www.kcwind.com/09.07.07.aspx |
| LM Glasfiber | Little Rock, Arkansas | 1100 | $150,000,000 | http://www.littlerock.org/citymanager/divisions/publicrelations/MediaReleases.aspx?ID=184 |
| Molded Fiber Glass Companies | Aberdeen, South Dakota | 750 | $40,000,000 | http://www.adcsd.com/press/MoldedFiberGlass.pdf |
| Nordex AG | Jonesboro, Arkansas | 700 | $100,000,000 | http://www.lucintel.com/newspage.aspx?sno=5400 |
| Nordic Windpower | Pocatello, Idaho | 160 | | http://www.reuters.com/article/pressRelease/idUS145972+10-Mar-2008+PRN20080310 |
| Polymarin Composites USA. | Little Rock, Arkansas | 830 | $20,000,000 | http://climate.weather.com/articles/turbines100801.html?page=1 |
| PPG Industries | Shelby, North Carolina | E-glass production with existing labor | $20,000,000 | http://www.manufacturing.net/wind-power-generates-n-c-jobs.aspx?menuid= |
| Siemens AG | Fort Madison, Iowa | 400 | $25,000,000 | http://w1.siemens.com/press/en/pr_cc/2007/09_sep/pg200709076_1463246.htm |
| Suzlon Rotor Corporation | Pipestone, Minnesota | 270 | $14,000,000 | http://minnesota.publicradio.org/display/web/2007/10/18/indiablades/ |
| TPI Composites | Newton, Iowa | 800 | $27,200,000 | http://www.renewableenergyworld.com/rea/news/story?id=53594<br>http://www.desmoinesregister.com/article/20081014/BUSINESS/8101402 |
| Vestas Wind Systems A/S | Brighton, Colorado | 1350 | $290,000,000 | http://www.colorado.gov/cs/Satellite/GovRitter/GOVR/1218795720308 |
| | Windsor, Colorado | 650 | $65,000,000 | http://www.rockymountainnews.com/news/2008/mar/05/wind-plant-officially-opens-doors/ |
| | Totals: | **8145** | **$849,200,000** | |

A research study undertaken by the National Renewable Energy Laboratory (NREL) states that the U.S will need between 50,000 and 500,000 new wind turbines within ten years (from 2020 - 2030) to accomplish its objectives of required wind power penetration (Greene, 2008).

Although wind energy market is emerging steadily in the U.S, currently it depends on overseas companies for the building of turbine blades. Manufacturers of huge, customized yachts are reportedly transitioning to composite wind turbine blades. An example is a renowned German company known as Abeking & Rasmussen which started building composite turbine blades in 1980 and is presently a leading manufacturing company in Germany. TPI Composites is an American yacht sailing company who first used resin infusion now frequently employed to mold blades of wind turbines. TPI is now a leading company providing its products to military and wind energy industry and have manufacturing facilities located in Rhode Island, Mexico, Iowa, China, Ohio, and China.

Another example is Knight & Carver; a yacht building company, whose experiences with large composite materials led them to repair a wind turbine blade. Since then, this company has rebuilt, repaired and manufactured hundreds and thousands of blades with facilities located in South Dakota and California.

## Wind Turbine Housings

A generator is used to convert the mechanical energy of turbine blades into equivalent electrical energy. These generators are in special cases known as the housing of generator and normally placed on the top of huge towers. These housings must be light in weight, low maintenance requirement, and aerodynamically well designed. Composite materials have been in use to build these housings.

Structurally, composite turbine housings are less challenging as compared to the blade assemblies. Thus, fabricators having experience in cost effective and high volume production of components can easily manufacture them. This provides an added advantage of the availability of opportunities in the renewable sector; to the whole industry and not just the elite, high standard fabricators. However, to produce low-cost turbine assemblies on an international level, several kinds of manufacturing processes are needed to be employed simultaneously.

**Figure 47:** Offshore wind turbines with aerodynamic turbine housing

## Wind Towers

Taller wind towers are designed to extract maximum energy from the wind due to their optimal height. The higher efficiency of taller wind towers results from the fact that wind speed increases as we get away from the earth's surface (ground level). Conventional wind towers have been built from steel products. However, Enercon constructed a 131-meter long wind turbine from 36 concrete segments.

Typically, composite materials are comparatively expensive than conventional materials (steel and concrete), but their robust processing and higher structural efficiencies are making them excellent candidates for high-performance structures. Considering the flexible manufacturing technology of composite materials, it is, now, possible to manufacture large structures on site, i.e. on-site manufacturing and installation of large poles and storage tanks. Modern developments in composite manufacturing and processing have led to the manufacturing and testing of giant composite structures, which were only fabricated with concrete and steel in the past. However, these modern structures possess superior mechanical and environmental characteristics, unlike conventional materials. Additionally, composite materials have revolutionized the repair mechanism of structural components. Deteriorated wind towers can easily be repaired with composite fibers sheets to strengthen the structure for bearing further loads. In near future, the composite industry will prove tremendously expedient in supporting the civil infrastructure of the world. Moreover, it can be predicted safely that spin-off composite technology and advanced supply chain techniques will revolutionize the renewable energy sector as well.

Composite materials have made it possible to develop offshore wind farms which are really hard to realize with metallic materials due to intense corrosive environment. Moreover, the maintenance of conventional wind tower is costly while composites are resistant to corrosion and require minimal repair costs. Composites are lightweight materials with high strength to weight ratio, which makes them the excellent choice for development of offshore floating wind towers. Composite wind turbines are also suitable for the development of small wind markets, for providing efficient energy to rural and remote areas. Maintenance free composite wind turbines are quite helpful in proving for cost-effective energy solutions.

## Water Turbines

Water, 1000 times denser than air, is one of the primary and efficient sources of energy. Marine engineers have long been dreaming of harnessing the power of ocean currents and tidal currents. Moving water has a massive potential to generate power, but the velocity of water is normally far than air velocity at wind farms. However, energy extracted from water sources is still comparatively higher than the power generated from other sources. It has always been a challenge to fabricate a device which can sustain the severe environment, and generate the power efficiently. A prototype of water-turbine and conceptual ideas related to modern water turbines are shown in Fig. 48. A power plant, called Verdant Power, was installed in East River, New York to address the issue of low velocities of water currents. It was observed that the plant collapsed in its first attempt to function in acute conditions due to the high speed of moving water currents.

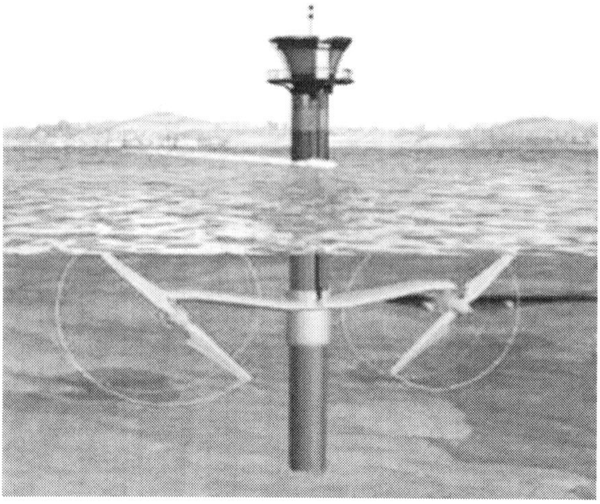

**Figure 48:** Underwater Turbines for energy generation at Florida

Tidal energy is considered more reliable than solar or wind energy due to the predictable nature of water currents for a certain location. Cyclic water currents are produced in tidal areas while constant water currents are produced by in marine environments. Maintenance of underwater structures is a more tedious job than that of offshore wind towers. Underwater installations require highly efficient material structures which can withstand severe atmospheric conditions and maintain the necessary mechanical strength. Composite materials have emerged as viable candidates for such applications. Similar to the multiple installations of wind turbines for generating high power at economical rates, water turbines also need to be installed in bulk quantity to harness the tidal energy efficiently. Composite materials are providing efficient yet cost effective solutions for the construction of turbine components. It is high time to expand the energy in developing high performance and cost saving composite materials for renewable energy production.

## Ocean Wave Energy

People having a visual or practical experience of offshore rides in a boat can estimate the amount of energy possessed by ocean waves. Ocean tides are an excellent source of energy, but its conversion into other forms of energy is a challenging job. Typically, the kinetic energy of a wave is converted into mechanical energy which is further converted into electrical power. According to a survey, it was reported that conversion of 20% wave energy at 50% efficiency would result in the energy equal to total hydro-power generated (conventionally) by U.S. in 2003.The harnessing of maximum offshore wave energy can contribute up to 20% of the total electrical energy production by U.S. (Robinson, 2008)."Point Capture" devices are used to extract the wave energy. Offshore buoys are quite expensive to maintain due to their enhanced susceptibility to degradation by the marine environment. However, the buoys manufactured from composite materials can serve as highly efficient and corrosion resistant structures. The construction of buoys and other offshore energy extraction devices from composite materials are still in the prototype phase. The shift from conventional to composite structures will unravel the horizons for manufacturing of novel energy capturing devices.

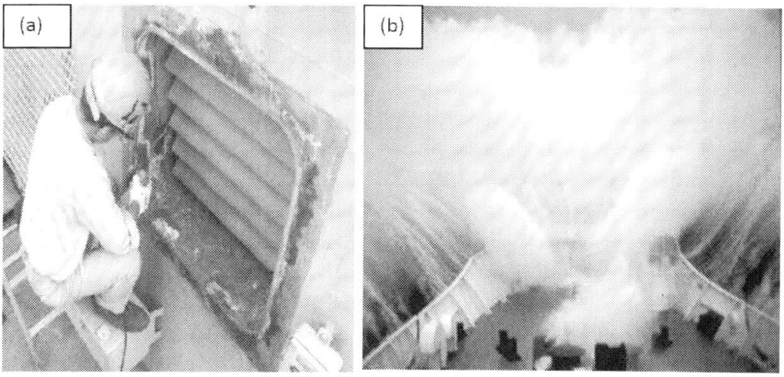


**Figure 49:** Ocean Wave Energy (a) Corrosion of ships due to ocean waves (b) Energetic waves reaching the ship surface

## POLYMER NANOCOMPOSITES (PNCS) FOR ENERGY APPLICATIONS

Polymer Nanocomposites have been studied extensively for multifarious applications including, environmental remediation, electromagnetic absorption, energy storage, actuation, sensing, information industry, transportation, defense system and novel catalysts (Cao et al., 2010; Tao et al., 2013; Moon et al., 2011; Wang et al., 2013).

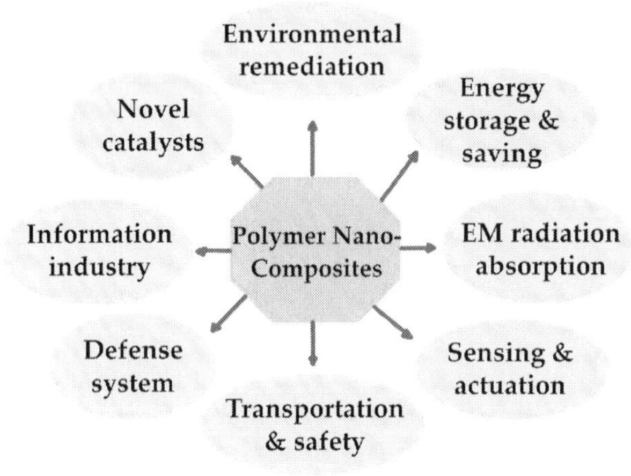


**Figure 50:** Applications of Polymer Nanocomposites (PNCs)

The use of PNCs in energy saving and energy storage applications has been illustrated in the subsequent text.

## Polymer Nanocomposites for Energy Storage Applications

The development of renewable energy storage mechanism, from abundantly available renewable energy resources such as the wind, water, and solar radiation, is crucial to combat the ever increasing energy demands in the world. Fossil energy resources (coal, oil, and gas) are depleting rapidly due to the increased usage in domestic and commercial applications. Therefore, there is dire need to store the energy obtained from natural renewable resources (Simon & Gogotsi, 2008; Wang et al., 2012; Díaz-González et al., 2012; Cook et al., 2010). Currently, electrochemical energy storage systems such as capacitors, fuel cells, and batteries have gained tremendous importance over other storage systems such as compressed air energy storage, thermal energy storage and pumped hydro storage (Díaz-González et al., 2012; Cao et al., 2015; Wang et al., 2015).

Polymer Nanocomposites are widely used in Lithium-Ion batteries (Ibrahim et al., 2008; Manthiram et al., 2008). Due to the benefits of high charge capacity, long life cycle and high working voltage, Lithium-ion batteries (LIBs) are widely used rechargeable batteries these days.

LIBs contain graphite, metal oxides and Si as the anode materials while $LiFePO_4$, LiCoO2, $LiMn_2O_4$, $Li[MnNiCo]_{O2}$, Li $[NiCoAl]O_2$ as the cathode materials (Ding et al., 2014). Inherently conductive polymers (ICPs) have attracted huge research endeavors to enhance the performance of traditional LIB terminal electrode materials (Etacheri et al., 2011). The ICPs contain organic electroactive functional groups which possess inherently rapid redox reactions than traditional inorganic electrode materials (Nyström et al., 2009). Moreover, flexible polymeric chains in ICPs are more stable than the crystalline particles during the charging-discharging process.

The ICPs coatings produced by in-situ polymerization technique are normally used to form conductive polymer structure in the polymer Nanocomposites based LIB electrodes. For instance, PNCs based anode materials (Si@PANI, SnO2@PPy, and CuO@PPy), and cathode materials (PPy@LiFeO2) show excellent performance due to rapid ionic and electronic exchange channels and enhanced mechanical integrity (Yin et al., 2012; Cui et al., 2011; Wu et al., 2013; Zhang et al., 2013).

Chao et al. (2014) have worked on highly integrated Polymer Nanocomposite based LIB cathode called «all-in-one» which is formed by the combination of carbon nanotubes and ICPs. The V2O5 based Nano-belt arrays (NBAs) were grown on 3D ultrathin graphite foam (UGF) followed by a subsequent coating of mesoporous thin film of a conducting polymer called poly(3,4-ethylenedioxythiophene- PEDOT) as shown in Fig. 51. Vanadium Pentoxide (V2O5) has a high current capacity (294 mAh/g) in a voltage range of 2.0-4.0 V was used as a host for reversible Li+ ion conversion reactions. UGF served as the current collector and the lightweight framework for the V2O5 NBAs growth. A homogeneous conductive film (15 nm thick) of PEDOT was specifically developed for the facilitation of electron transfer along with retention of NBAs structure during long term cycles. The addition of PEDOT causes the decrement in electrode polarization and prolongs the discharge time at high voltage ranges (above 3 V), which ultimately results in high energy density and increased high voltage capacity. The improvement in the capacity of the battery by the addition of PEDOT was affirmed by the charge-discharge peaks in Fig. 51d. Apparently, the PEDOT/UGF−V2O5 shell /core electrode showed superior charge-discharge characteristics than the UGF−V2O5 specimen without PEDOT films. Moreover,

PEDOT/UGF−V2O5 specimen exhibited high specific capacity profiles as compared to simple (UGF−V2O5) specimen, i.e. Specific capacities ranging between 297 mA h/g at 1 C and 115 mA h/g at 80 0C (Fig. 51e). Charge transfer resistance (Rct) also showed decrement as shown in electrical impedance spectroscopy data in Fig. 51f.

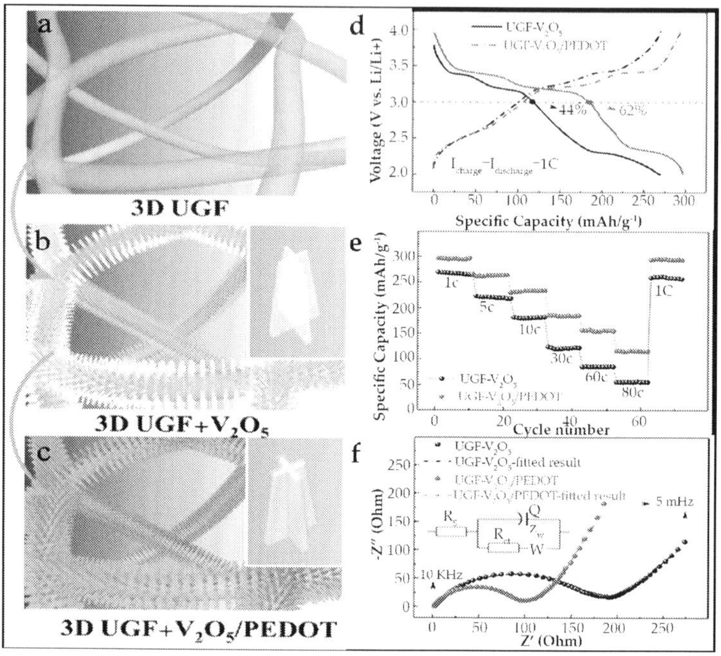


**Figure 51:** a–c) Schematics of the blend of UGF−V2O5/PEDOT shell terminal. (d) Charge–discharge bends (e) Rate capacity (f) Impedance spectroscopy plots at charged stage.

Other than Lithium ion batteries, the Lithium sulfide (Li/S) batteries have also attracted considerable attention as energy storage systems in modern power storage applications. TLi/S batteries exhibit three folds higher theoretical energy densities (around 2600 W h kg-1) than conventional LIBs (Bruce et al., 2012). However, Li/S batteries undergo expansion due to the presence of sulfur which results in lower efficiencies, i.e. low electronic/ionic conductivity of both lithium sulfide and sulfur (Huang et al., 2015). Numerous experiments have been carried out to replace conventional carbon matrix with modern ICPs such as PEDOT, PANI, and PPy. The ICP/sulfur Nanocomposites form an intricate conductive structure, and an exclusive soft porous and soft structure which is quite lucrative for improving the long-term performance of the cell.

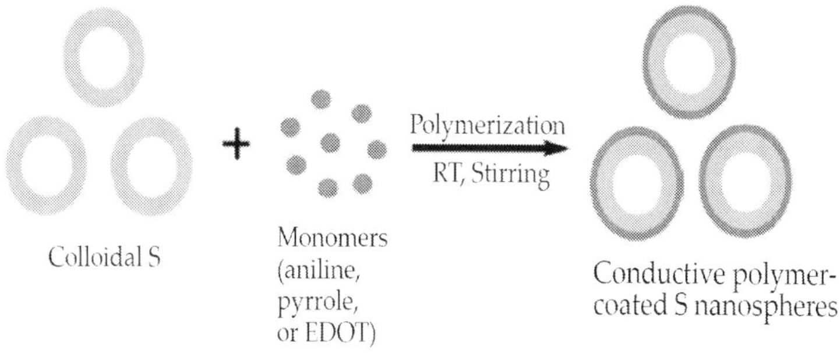


Figure 52: Schematic of the manufacturing process of conductive polymer-coated sulfur Nano-spheres.

## Polymer Nanocomposites for Energy Saving Applications

The concept of energy saving is based on human ability to reduce the fossil fuel consumption, improve the economy of the specific country or save the environment from harmful effects of electrical appliances. According to the reports, around 35-40% energy is being utilized for lightening, cooling, heating and ventilation of buildings. Air conditioners (ACs) are normally used to maintain and control the temperature in the buildings. It is, therefore, imperative to devise a mechanism for controlling the heat transfer between exterior and interior of the buildings. Presently, smart windows are being installed in the buildings to control the temperature in the building. Smart windows automatically increase the heat input in cold weathers and decrease the thermal heat input in hot weathers. Smart windows are ultimately reducing the energy consumption in the domestic and commercial buildings. Smart chromic materials are an excellent choice for coating application of building windows (Ferrara & Bengisu, 2013). Based on the triggering mechanism for windows' modulation, there are four different types of smart chromic materials which include photochromic-PC (activated by light), gasochromic-GC (triggered by hydrogen gas), electro-chromic-EC (activated by the electricity), and thermochromic-TC (stimulated by heat) (Gao et al., 2012).

Experiments have shown that the electrochromic windows exhibit superior controllability than TC or photochromic windows without any requirement of trapped gas as in the case of GC windows. Therefore, EC windows are being used extensively due to their promising characteristics (Jelle et al., 2012). Polymer Nanocomposites are versatile materials for fabrication of EC coatings on windows.

Electrochromism is an optical change which occurs upon oxidation (electron gain) or reduction (electron loss) after the passage of electrical current through a suitable material.

The phenomenon of Electrochromism is under comprehensive investigation for manufacturing controllable light transmission or light reflection devices to store electrical and optical energy. The novel devices are manufactured from Nanocomposites which are extensively used in anti-glare mirrors of vehicles, protective eyewear, glare-reduction glass for commercial and private buildings, and 'smart windows' for both buildings and cars. Electrochromism is an effective phenomenon which is used to reduce the energy consumption by controlling heating and cooling rates of buildings, eventually, saving the energy.

The ICPs experience the Electrochromism due to the alterations in $\pi$-electronic characteristics followed by reversible extraction and insertion of ions through the polymeric films upon reduction and oxidation reactions. Electrochromic characteristics of ICPs have attained immense attention in both industry and academia due to their robust processing, higher optical contrast, quick response time and capability to tailor their structure for generation of multicolor electro-chromes. Shrinkage of ICP thin films during the cyclic redox reactions poses a threat to their mechanical and electrochemical integrity for commercial Electrochromic applications. Wei et al (2013) addressed the issue by incorporating graphite oxide (GONPs) and Tungsten oxide (WO3) nanoparticles in the polyaniline (PANI) matrix. The interactions of polymeric matrix and nanoparticles assisted in the reduction of mechanical stresses, ultimately, providing improved stability to the Electrochromic structure. Nanocomposite films are fabricated by spin coating of the Nano-fillers on a conductive transparent substrate (Indium Tin Oxide-ITO). The grown Nano-fillers act as catalysts for PANI growth through electropolymerization method. Nanocomposite thin films exhibit multicolor Electrochromism by showing different light colors, i.e. light yellow color at -0.2 V (reduced condition), light green color at 0.5 V and 0.8 V (partially oxidized form), and dark blue color at 1.0 V (completely oxidized state), in an aqueous solution comprising of 0.5 M H2SO4 (Fig. 53).

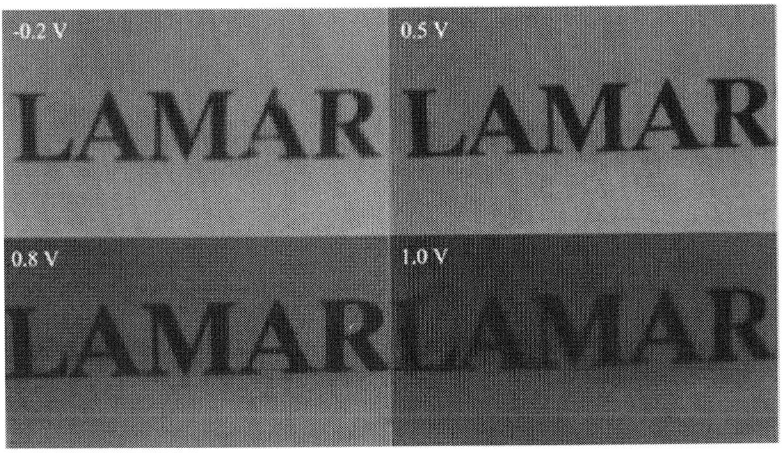


**Figure 53:** Electrochromic appeared in the PANI/GO Nanocomposites thin films.

## RECENT DEVELOPMENTS IN NANOCOMPOSITES

Currently, high-performance energy storage capacitors are being fabricated from Nanomaterials. Dielectric Nanocomposites are typically manufactured by blending inorganic nanoparticles (NPs) into polymer resins. Presently, capacitors based on polyvinylidene fluoride are being extensively manufactured with incorporation of multifarious NPs such as $BaTiO_3$, $BaZr_{1-x}Ti_xO_3$ [BZT], $Ba_{0.5}Sr_{0.5}TiO_3$ [BST] and $Ba_{1-x}Ca_xTiO_3$ [BCT] (Adireddy et al., 2014). Colloidal NPs are usually manufactured in high quantity through a solvothermal process and utilized as reinforcing fillers in polyvinylidene fluoride PVDF polymer resin. The dielectric polymer thin films exhibit higher electrical and electronic breakdown strength, superior dielectric constants, and improved dielectric energy density (Table 1). Nanocomposites show greater values of dielectric permittivity, breakdown strength, and versatility than that of the polymeric matrix alone. Nanocomposites have replaced traditional ceramic materials for the capacitor and other electronic applications, due to their lighter weight and enhanced flexibility (Liu & Zhai, 2014). Usually, Nano-particles are not distributed evenly in polymers. This nonhomogeneity affects dielectric and breakdown properties of polymers negatively, and therefore researchers are focusing their research on investigating the methods for proper dispersion of Nanoparticles. It is quite possible to achieve high uniformity in composition of composites due to surface fictionalization properties of Nanoparticles. Moreover, it is also possible to achieve higher breakdown strength in composites manufactured by Nanoparticles. For instance, $BaTiO_3$was modified with Polyvinylpyrrolidone (PVP) to form a more homogeneous product. The resultant product had a higher dielectric permittivity of about 77 at 1 kHz, which is roughly 28% more than BaTiO3 (Masuduzzaman et al., 2012).

Nanomaterials have also been modified for another purpose; that is for investigation of interface properties. An example is the modification of $BaTiO_3$with DN-101. As a result of this modification, polarization response of the composite was increased to 240 MV/m which is approximately an improvement of 18%. The increase in dielectric strength of the material was also observed from 240 MV/m to 260 MV/m. This is equivalent to an increase of 6.5 to 9 J/cm3 in its energy density (Gao et al., 2014).

Recently, modifications have been made between matrix and particles to lessen the damage to breakdown strength which occurs due to the concentration of charges and low breakdown interfaces (Puli et al., 2014; Zhou et al., 2011). The main objective is to create a bond between a polymer and the particle so as to exclude any rushed transition from polymer to ceramic. Hydroxyl groups have been applied to the surface of $BaTiO_3$by Zhou et al (Zhou et al., 2011). Fluorine creates hydrogen bonds with hydroxyl groups on mixing with PVDF. This modification allows the larger volume to be mixed which resultantly, increases breakdown percentage by 17%.

Another technique which is used to increase the contact between the particle and matrix is by grafting the matrix material of interest with Nanoparticle (Wang et al., 2011). To demonstrate this method, long 240 kDa PVDF copolymer chains were grown onto TiO2 Nano-particles. After this, it was easy to disperse the particles in the form of a polymer matrix and cast it into thin films afterward (Riggs et al., 2015). Breakdown field of the modified particles was slightly decreased. On the other hand, the dielectric constant was increased following logarithmical laws of mixing. An ultimate method is the direct binding of Nanoparticles with PVDF matrix (Dou et al., 2009). Experiments show that ferroelectric Nanoparticles are functional with titanate coupling agent (NDZ101). This material has a functional alloy which can react with PVDF copolymer to increase the binding strength between the particle and the matrix to improve the breakdown properties.

## FUTURE PROSPECTS

Composite manufacturing facilities of the United States are currently the most innovative and productive in the world. Unfortunately, we are facing a decrease in the demand for marine and construction related composite materials which are leaving us with extra manufacturing capacity. Use of composites in the renewable energy sector is still somewhat new and in its design or prototype phase. Europe is currently leading the production of wind turbine blades. Focus, therefore, should be in both renewable energy sector and wind turbine technology if the U.S wants to become a world leader in the composite industry. For this purpose, the focus should be placed on market development and R & D of composites.

Below are some initiatives which can lead us onto the path of energy independence by creating environment-friendly jobs out of a rapidly declining industry.

- A precise evaluation of prevailing composite manufacturing competencies (that can aid renewable energy market) of the United States needs to be developed.
- Experts of the composite industry should collaborate and interact with renewable energy industries to convince them to use the composite structures by presenting related technical and business data.
- Although the U.S is considered a world leader in transferring the technology from educational institutes and laboratories to the commercial market, still, there needs to be a government entity which must ensure that funded efforts provided are in accordance with national energy and economic objectives.
- The Defense Advanced Research Projects Agency (DARPA) has presented a practical model which shows the successful means to fund innovative companies having massive repayment potential. Small, innovative companies based on DARPA model and working in the renewable energy sector

should be provided basic funds. Funds should be provided to both small and large companies on the non-matching basis so that they may not rule out small industries.

By spending a small government fund on the development of composites and renewable energy industries, a favorable situation can be created which will benefit both consumers and workers.

## REFERENCES


1. Adireddy, S., Puli, V. S., Sklare, S. C., Elupula, R., Lou, T. J., Grayson, S., & Chrisey, D. B. (2014). Flexible ferroelectric polymer-nanocrystal composite films for large-scale capacitive energy storage. PRINCIPLES, STRUCTURE AND APPLICATIONS, 55.
2. Adireddy, S., Puli, V. S., Sklare, S. C., Lou, T. J., Riggs, B. C., Elupula, R., & Chrisey, D. B. (2014). PVDF–BaSrTiO3 nanocomposites for flexible electrical energy storage devices. Emerging Materials Research, 3(6), 265-270.
3. Bruce, P. G., Freunberger, S. A., Hardwick, L. J., & Tarascon, J. M. (2012). Li-O2 and Li-S batteries with high energy storage. Nature materials, 11(1), 19-29.
4. Cao, H., Wang, X., Gu, H., Liu, J., Luan, L., Liu, W., & Guo, Z. (2015). Carbon coated manganese monoxide octahedron negative-electrode for lithium-ion batteries with enhanced performance. RSC Advances, 5(44), 34566-34571.
5. Cao, A., Liu, Z., Chu, S., Wu, M., Ye, Z., Cai, Z. & Liu, Y. (2010). A Facile One-step Method to Produce Graphene–CdS Quantum Dot Nanocomposites as Promising Optoelectronic Materials. Advanced materials, 22(1), 103-106.
6. Chao, D., Xia, X., Liu, J., Fan, Z., Ng, C. F., Lin, J. & Fan, H. J. (2014). A V2O5/conductive-polymer core/shell nanobelt array on three-dimensional graphite foam: A high-rate, ultrastable, and freestanding cathode for lithium-ion batteries. Advanced Materials, 26(33), 5794-5800.
7. Cook, T. R., Dogutan, D. K., Reece, S. Y., Surendranath, Y., Teets, T. S., & Nocera, D. G. (2010). Solar energy supply and storage for the legacy and nonlegacy worlds. Chemical reviews, 110(11), 6474-

6502.

8. Cui, L., Shen, J., Cheng, F., Tao, Z., & Chen, J. (2011). SnO 2 nanoparticles@ polypyrrole nanowires composite as anode materials for rechargeable lithium-ion batteries. Journal of Power Sources, 196(4), 2195-2201.
9. Díaz-González, F., Sumper, A., Gomis-Bellmunt, O., & Villafáfila-Robles, R. (2012). A review of energy storage technologies for wind power applications. Renewable and sustainable energy reviews, 16(4), 2154-2171.
10. Díaz-González, F., Sumper, A., Gomis-Bellmunt, O., & Villafáfila-Robles, R. (2012). A review of energy storage technologies for wind power applications. Renewable and sustainable energy reviews, 16(4), 2154-217.
11. Ding, K., Gu, H., Zheng, C., Liu, L., Liu, L., Yan, X., & Guo, Z. (2014). Octagonal prism shaped lithium iron phosphate composite particles as positive electrode materials for rechargeable lithium-ion battery. Electrochimica Acta, 146, 585-590.
12. Dou, X., Liu, X., Zhang, Y., Feng, H., Chen, J. F., & Du, S. (2009). Improved dielectric strength of barium titanate-polyvinylidene fluoride nanocomposite. Applied physics letters, 95(13), 132904.
13. Etacheri, V., Marom, R., Elazari, R., Salitra, G., & Aurbach, D. (2011). Challenges in the development of advanced Li-ion batteries: a review. Energy & Environmental Science, 4(9), 3243-3262.
14. Enercon, "Construction of world's most powerful wind turbines in progress in Emden," news 28/09/07, http://www.enercon.de/en/_home.htm
15. Ferrara, M., & Bengisu, M. (2013). Materials that change color: smart materials, intelligent design. Springer Science & Business Media.
16. Gao, L., He, J., Hu, J., & Li, Y. (2014). Large enhancement in polarization response and energy storage properties of poly (vinylidene fluoride) by improving the interface effect in nanocomposites. The Journal of Physical Chemistry C, 118(2), 831-838
17. Greene, E. (2008). Composites for Renewable Energy. Obama-Biden Transition Website, December

18. Greene, E. (2008). Composites for Renewable Energy. Obama-Biden Transition Website, December.
19. Greene, E. (2008). Composites for Renewable Energy. Obama-Biden Transition Website, December
20. Huang, L., Cheng, J., Li, X., Yuan, D., Ni, W., Qu, G. & Wang, B. (2015). Sulfur quantum dots wrapped by conductive polymer shell with internal void spaces for high-performance lithium–sulfur batteries. Journal of Materials Chemistry A, 3(7), 4049-4057.
21. Ibrahim, H., Ilinca, A., & Perron, J. (2008). Energy storage systems—characteristics and comparisons. Renewable and sustainable energy reviews, 12(5), 1221-1250.
22. Jelle, B. P., Hynd, A., Gustavsen, A., Arasteh, D., Goudey, H., & Hart, R. (2012). Fenestration of today and tomorrow: A state-of-the-art review and future research opportunities. Solar Energy Materials and Solar Cells, 96, 1-28.
23. Kao, J., Thorkelsson, K., Bai, P., Rancatore, B. J., & Xu, T. (2013). Toward functional nanocomposites: taking the best of nanoparticles, polymers, and small molecules. Chemical Society Reviews, 42(7), 2654-2678.
24. Kokate, M., Garadkar, K., & Gole, A. (2013). One pot synthesis of magnetite–silica nanocomposites: applications as tags, entrapment matrix and in water purification. Journal of Materials Chemistry A, 1(6), 2022-2029.
25. Kumar, S. K., Jouault, N., Benicewicz, B., & Neely, T. (2013). Nanocomposites with polymer grafted nanoparticles. Macromolecules, 46(9), 3199-3214.
26. Laxson, A. S., Hand, M. M., & Blair, N. (2006). High wind penetration impact on us wind manufacturing capacity and critical resources. National Renewable Energy Laboratory.
27. Li, Z. F., Zhang, H., Liu, Q., Sun, L., Stanciu, L., & Xie, J. (2013). Fabrication of high-surface-area graphene/polyaniline nanocomposites and their application in supercapacitors. ACS applied materials & interfaces, 5(7), 2685-2691.
28. Liu, S., & Zhai, J. (2014). A small loading of surface-modified Ba 0.6 Sr 0.4 TiO 3 nanofiber-filled nanocomposites with enhanced dielectric constant and energy density. RSC Advances, 4(77),

40973-40979.

29. Manthiram, A., Murugan, A. V., Sarkar, A., & Muraliganth, T. (2008). Nanostructured electrode materials for electrochemical energy storage and conversion. Energy & Environmental Science, 1(6), 621-638.
30. Masuduzzaman, M., Xie, S., Chung, J., Varghese, D., Rodriguez, J., Krishnan, S., & Ashraful Alam, M. (2012). The origin of broad distribution of breakdown times in polycrystalline thin film dielectrics. Applied Physics Letters, 101(15), 153511
31. Moon, R. J., Martini, A., Nairn, J., Simonsen, J., & Youngblood, J. (2011). Cellulose nanomaterials review: structure, properties and nanocomposites. Chemical Society Reviews, 40(7), 3941-3994.
32. Mortimer, R. J., Dyer, A. L., & Reynolds, J. R. (2006). Electrochromic organic and polymeric materials for display applications. Displays, 27(1), 2-18.
33. Nyström, G., Razaq, A., Strømme, M., Nyholm, L., & Mihranyan, A. (2009). Ultrafast all-polymer paper-based batteries. Nano letters, 9(10), 3635-3639.
34. Puli, V. S., Elupula, R., Riggs, B. C., Grayson, S. M., Katiyar, R. S., & Chrisey, D. B. (2014). Surface modified BaTiO 3-polystyrene nanocomposites for energy storage. International Journal of Nanotechnology, 11(9-1011), 910-920.
35. Pérez-Juste, J., Pastoriza-Santos, I., & Liz-Marzán, L. M. (2013). Multifunctionality in metal@ microgel colloidal nanocomposites. Journal of Materials Chemistry A, 1(1), 20-26.
36. Red, C. (2008). Wind turbine blades: Big and getting bigger. Composites technology, 6, 42-47.
37. Riggs, B. C., Adireddy, S., Rehm, C. H., Puli, V. S., Elupula, R., & Chrisey, D. B. (2015). Polymer Nanocomposites for Energy Storage Applications. Materials Today: Proceedings, 2(6), 3853-3863.
38. Robinson, M. (2006). Renewable energy technologies for use on the outer continental shelf. National Renewable Energy Laboratory, June, 6.
39. Simon, P., & Gogotsi, Y. (2008). Materials for electrochemical capacitors. Nature materials, 7(11), 845-854.

40. Shin, H., Kim, Y., Bhuvana, T., Lee, J., Yang, X., Park, C., & Kim, E. (2011). Color combination of conductive polymers for black electrochromism. ACS applied materials & interfaces, 4(1), 185-191.
41. Sroog, C. E. (1991). Polyimides. Progress in Polymer Science, 16(4), 561-694.
42. Tao, P., Viswanath, A., Li, Y., Siegel, R. W., Benicewicz, B. C., & Schadler, L. S. (2013). Bulk transparent epoxy nanocomposites filled with poly (glycidyl methacrylate) brush-grafted TiO 2 nanoparticles. Polymer, 54(6), 1639-1646.
43. Wang, L., Wang, D., Dong, Z., Zhang, F., & Jin, J. (2013). Interface chemistry engineering for stable cycling of reduced GO/SnO2 nanocomposites for lithium ion battery. Nano letters, 13(4), 1711-1716.
44. Wang, G., Zhang, L., & Zhang, J. (2012). A review of electrode materials for electrochemical supercapacitors. Chemical Society Reviews, 41(2), 797-828.
45. Wang, Y., He, Q., Ding, K., Wei, H., Guo, J., Wang, Q. & Wei, S. (2015). Multiwalled Carbon Nanotubes Composited with Palladium Nanocatalysts for Highly Efficient Ethanol Oxidation. Journal of The Electrochemical Society, 162(7), F755-F763
46. Wang, L., Goodloe, G. W., Stallman, B. J., & Cammarata, V. (1996). Synthesis, electrooxidation, and characterization of bis (diphenylamine) naphthalene diimide. Chemistry of materials, 8(6), 1175-1181.
47. Wang, J. Electrochem. Soc., 1998, 145, 2648–2654.
48. Wang, Q., & Zhu, L. (2011). Polymer nanocomposites for electrical energy storage. Journal of Polymer Science Part B: Polymer Physics, 49(20), 1421-1429.
49. Wei, S., Wang, Q., Zhu, J., Sun, L., Lin, H., & Guo, Z. (2011). Multifunctional composite core–shell nanoparticles. Nanoscale, 3(11), 4474-4502.
50. Wei, H., He, C., Liu, J., Gu, H., Wang, Y., Yan, X. & Wei, S. (2015). Electropolymerized polypyrrole nanocomposites with cobalt oxide coated on carbon paper for electrochemical energy storage. Polymer, 67, 192-199.

51. Wei, H., Gu, H., Guo, J., Wei, S., & Guo, Z. (2013). Electropolymerized polyaniline nanocomposites from multi-walled carbon nanotubes with tuned surface functionalities for electrochemical energy storage. Journal of The Electrochemical Society, 160(7), G3038-G3045.
52. Wei, H., Yan, X., Wu, S., Luo, Z., Wei, S., & Guo, Z. (2012). Electropolymerized polyaniline stabilized tungsten oxide nanocomposite films: electrochromic behavior and electrochemical energy storage. The Journal of Physical Chemistry C, 116(47), 25052-25064.
53. Wei, H., Yan, X., Li, Y., Wu, S., Wang, A., Wei, S., & Guo, Z. (2012). Hybrid electrochromic fluorescent poly (DNTD)/CdSe@ZnS composite films. The Journal of Physical Chemistry C, 116(7), 4500-4510.
54. Wei, H., Yan, X., Li, Y., Gu, H., Wu, S., Ding, K., & Guo, Z. (2012). Electrochromic poly (DNTD)/WO3 nanocomposite films via electorpolymerization. The Journal of Physical Chemistry C, 116(30), 16286-16293
55. Wiser, R., Bollinger, M., Barbose, G., Belyeu, K., Hand, M., Heimiller, D., & Musial, W. (2008). Annual report on US wind power installation, cost, and performance trends: 2006. Lawrence Berkeley National Laboratory.
56. Wu, H., Yu, G., Pan, L., Liu, N., McDowell, M. T., Bao, Z., & Cui, Y. (2013). Stable Li-ion battery anodes by in-situ polymerization of conducting hydrogel to conformally coat silicon nanoparticles. Nature communications, 4.
57. Yang, C., Wei, H., Guan, L., Guo, J., Wang, Y., Yan, X., & Guo, Z. (2015). Polymer nanocomposites for energy storage, energy saving, and anticorrosion. Journal of Materials Chemistry A, 3(29), 14929-14941.
58. Yu, K., Niu, Y., Zhou, Y., Bai, Y., & Wang, H. (2013). Nanocomposites of Surface-Modified BaTiO3 Nanoparticles Filled Ferroelectric Polymer with Enhanced Energy Density. Journal of the American Ceramic Society, 96(8), 2519-2524.
59. Yin, Z., Ding, Y., Zheng, Q., & Guan, L. (2012). CuO/polypyrrole core–shell nanocomposites as anode materials for lithium-ion batteries. Electrochemistry Communications, 20, 40-43.

60. Zhang, Z. J., Wang, J. Z., Chou, S. L., Liu, H. K., Ozawa, K., & Li, H. J. (2013). Polypyrrole-coated α-LiFeO 2 nanocomposite with enhanced electrochemical properties for lithium-ion batteries. Electrochimica Acta, 108, 820-826.

61. Zhou, T., Zha, J. W., Cui, R. Y., Fan, B. H., Yuan, J. K., & Dang, Z. M. (2011). Improving dielectric properties of BaTiO3/ferroelectric polymer composites by employing surface hydroxylated BaTiO3 nanoparticles. ACS applied materials & interfaces, 3(7), 2184-2188.

# CHAPTER 6

# AEROSPACE APPLICATIONS OF COMPOSITE MATERIALS

## BACKGROUND

The usage of composites in aerospace applications has evolved significantly since their first use in the 1970s. Advanced composite materials have been recognized as the essential materials for various aerospace and aeronautic applications due to their excellent mechanical properties (specific stiffness and specific strength). Composite materials were initially used for the development of military aircraft. Presently, composite materials are being utilized for the manufacturing of aircraft, airplanes, space shuttles, satellites and launch systems. The use of composites for manufacturing of commercial airplanes and aviation aircraft has increased significantly during the last decade (Baker, 2004). Moreover, there is a variety of air transport structures with the substantial use of composites, which are waiting for flight certification of Federal Aviation Administration (FAA) in the immediate future.

Composites are a versatile class of materials which are being used in structural aerospace applications including aircraft structures, spacecraft components, passenger airplane wings, space shuttles interiors and fighter plane components. Applications of composite materials range from entire aircraft structures (e.g. Beech Starship) to helicopter rotor blades, wing assemblies, instrument enclosures, and propellers. Despite the extensive use of composite materials in aerospace structures, composites technology is facing some serious challenges for their exploitation in the manufacturing of commercial aircraft structures, i.e. fuselage and wing. The primary challenges include fuel containment, damage tolerance, lightening protection, nondestructive inspection and repair, cost effective manufacturing and failure prediction. Comprehensive research and development work is required to address these issues, which is only possible by conducting a dedicated scientific research in the field of aerospace composites.

The modern composites technology is playing a cordial role to revolutionize the aerospace industry. With recent developments in materials science and engineering, there are various novel techniques which will possibly speed up the manufacturing of aerospace structural components including sensors, solar panels, fuselage, fuel tanks and launch vehicle motor casings. The new composite technology will exploit the novel processing techniques and materials to provide active flow control, intelligent materials and structures, durable material design, robust manufacturing technologies and practical use of nanotechnology. Additionally, the deliberate integration of these modern technologies will provide a basis for construction of high-performance aerospace structures. However, the successful inclusion of these novel technologies cannot be realized with conventional structural materials. Consequently, it is imperative to direct the efforts towards the development of next generation composite technology.

Distinct categories of composite materials are employed for fulfilling the specific requirements in the aerospace structures. Carbon fiber, for instance, has exclusive fatigue behavior but it acts as a brittle material under extreme loading conditions. Rolls-Royce discovered the brittleness of carbon fibers in the 1960's when the RB211 jet engine (containing carbon fiber compressor blades) crashed catastrophically due to bird strikes. On the contrary, aluminum structures have a known fatigue lifetime, but its specific strength and specific stiffness are comparatively small. Carbon fiber reinforced composites, however, provide significant fatigue and impact resistance due to the presence of both high strength and high toughness materials, i.e. carbon fibers and epoxy resins. Aramid fiber composites are another class of composite materials which are broadly used to manufacture honeycomb structures for the development of lightweight and stiff fuel tanks, floors and wing components (Asundi & Choi, 1997). Kevlar is a famous trademark of DuPont which is manufactured from aramid fibers. Many kinds of composite materials are used in aerospace industry, but carbon fiber composites are the most commonly used materials in structural aerospace applications (Mrazova, 2013).

It has been reported that Boeing successfully employed 1,500 composite components to substitute 11,000 metallic parts in a helicopter. Besides, the use of composite materials for repair and maintenance of commercial aircraft is gaining considerable attention these days.

## BENEFITS OF COMPOSITE MATERIALS IN AERO-SPACE INDUSTRY

There are numerous reasons to employ composites in aerospace applications. Some significant advantages of composite materials are illustrated in the following text.

- Composite materials are and an excellent choice for manufacturing structures with high strength to weight with approximately 20-50% reduction in the overall weight.
- It is quite convenient to assemble the complex composite parts using and rotational molding processes and automated layup machinery.
- Monocoque ('single-shell') composite structures provide higher specific strength at lower weights.
- Mechanical properties of the composite structures can be custom-made via 'lay-up' manufacturing design by maneuvering the thicknesses and orientation of cloth reinforcement.
- Composite materials exhibit high thermal stability which implies that they can withstand extreme temperature variations (90°F to -67°F from runway to 35,000 feet height for airplanes) without undergoing expansion or contraction.
- Composite materials have high impact resistance as compared

to conventional materials. For example, Kevlar armor shielded planes have minimized susceptibility to accidental damage of the engine pylons which contain fuel lines and engine controls.

- High damage tolerance of composite materials increases accident survivability of aerospace structures.
- Composites are excellent anticorrosion materials against galvanic corrosion, which makes them and the ideal choice for extreme weather condition (i.e. space).
- Composite materials display greater resistance to wear and fatigue as compared to metallic structures.

## HISTORY OF COMPOSITE APPLICATIONS IN AEROSPACE SECTOR

The applications of composites in the aerospace industry are pervasive these days since their first use in a military aircraft in the 1970's (Soutis, 2005). The evolution of composite technology has been materialized by the historical developments in the materials and processing techniques. The initial stage of the aerospace composite technology was led by the defense industry which resulted in the construction of the commercial rotorcrafts and aircraft. Origin of the advanced composite materials can be traced back to the invention of carbon fibers in Japan and boron fibers in the USA in the 1960s. The first aerospace application of composite materials was the manufacturing of the horizontal stabilizer of F14 aircraft in the 1970s, followed by composite manufacturing of F-15 and F-16 aircraft of the US Air Force. In the late 1970s, the commercial applications of composite materials were administered by the NASA ACEE Program.

Famous space manufacturing industries started the use of composite materials in the 1980s. Composite based horizontal stabilizer of Boeing 737 was manufactured and certified in 1982. Weight content of the composite components approached approximately 10% in the structures of Airbus 300 and 310, McDonnell-Douglas MD-82, MD-83 and MD-87 and Boeing 737, 757 and 767, during the late 1980's. The Airbus 320 was the pioneer commercial aircraft which experienced the use of more than 10% composite materials for the first time. By 2000, about 30% reduction in the structural weight of the US Air Force aircraft including

F-22 and B2 and the U.S. Navy aircraft, V-22. The aerospace industry is determined to develop military and commercial airframes with widespread composite usage in wings, fuel tanks, and fuselage. "Sonic Cruiser" of Boeing and 650-seater Airbus A380 are recent examples of the composite applications in aircraft. Currently, new business jets embedded with composite sandwich panels in fuselage structures are under consideration for the certification by Federal Aviation Administration. Propeller-driven aircraft have also integrated composites in their airframe structures. The use of composite components in the propeller-driven aircraft such as Cirrus Design Corp. SR 20 and Lancair LC40-550FG was certified in 1998 (Pipes, 2002). New rotorcraft buses incorporate composite materials in rotor drive systems, rotor blades, and airframes. Modern examples of composite manufactured rotorcrafts include the Bell Textron BA609 and the Sikorsky S92.

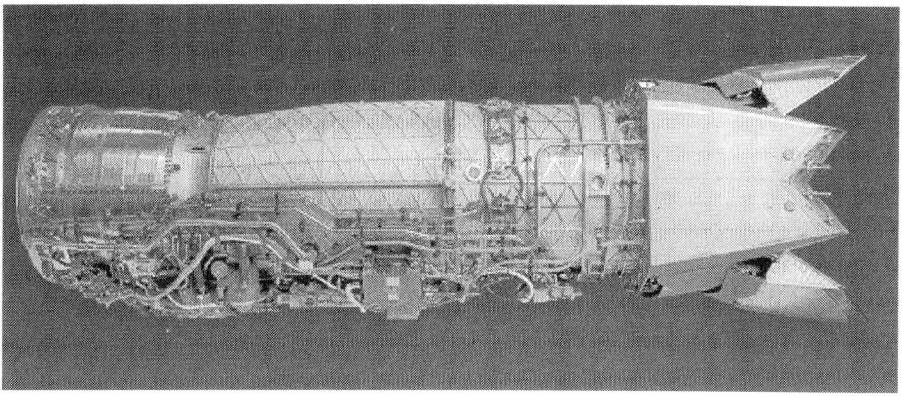

**Figure 54:** Fuselage of F22 Raptor made of composite materials

Space launch vehicles and satellite structures experience a broad range of environmental conditions during both the flight and service modes. Space environment presents some atmospheric factors which include space radiations, extreme temperatures, and cosmic waves. These factors can only be addressed by utilizing the advanced materials which can withstand extreme climatic condition. Composite materials are the most appropriate choice to combat the challenges of the space environment. Carbon fiber composites and graphite/epoxy composites are the most widely used materials for satellite applications. Intelsat IV was the first communication satellite which consisted of graphite/epoxy composite components. This composite design *was* opted to meet the requirements for low weight and small deflections (Mayer, 1974). In the 1990s the US space industry utilized carbon fiber composites for the production of Reusable Launch Vehicles (RLV).

**Figure 55:** Composite motor casings of launch vehicles

## CRITICAL FEATURES OF THE AEROSPACE STRUCTURES

Comprehensive details of essential requirements and their effects on aerospace structures are illustrated in Table 6.

**Table 6:** Salient features of an aerospace structure

| Requirement | Applicability | Effect |
|---|---|---|
| • Light-weight | All Aerosace Programmes | • Semi-monocoque construction<br>* Thin-walled-box or stiffened structures<br>• Use of low density materials:<br>* wood * Al-alloys * Composites<br>• High strength/weight, High stiffness/weight |
| • High reliability | All space programmes | • Strict quality control<br>• Extensive testing for reliable data<br>• Certification: Proof of design |
| • Passenger safety | Passenger vehicles | • Use of fire retardant materials<br>• Extensive testing: Crashworthiness |
| • Durability-Fatigue and corrosion Degradation: Vacuum Radiation Thermal | Aircraft<br>Spacecraft | • Extensive fatigue analysis/testing<br>* Al-allows do not have a fatigue limit<br>• Corrosion prevention schemes<br>• Issues of damae and sae-life, life extension<br>• Extensive testing for required environment<br>• Thin materials with high integrity |
| • Aerodynamic performance | Aircraft<br>Reusable spacecraft | • Highly complex loading<br>• Thin flexible wings and control surfaces<br>* Deformed shape-Aero elasticity * Dynamics<br>• Complex contoured shapes<br>* Manufacturability: N/C Machining; Moulding |
| • Multi-role or functionality | All Aerospace programmes | • Efficient design<br>• Use: composites with functional properties |
| • Fly-by-wire | Aircrafts, mostly for fighters but also some in passenger a/c | • Structure-control interactions<br>* Aero-servo-elasticity<br>• Extensive use of computers and electronics<br>* EMI shielding |
| • Stealth | Specific military aerospace applications | • Specific surface and shape of aircraft<br>* Stealth coatings |
| • All-Weather operation | Aircraft | • Lightning protection erosion resistance |

Apart from the features mentioned in Table 6, aerospace structures are designed to provide easy access to the maintenance tools and protect the fuel containers with a seal. Unique materials are required for constructing passenger carriages to fulfill the safety requirements such as crashworthiness and fire resistance (Perret et al., 2011). Features of the space environment (radiation, vacuum and thermal cycling) are also taken into account while designing an aerospace structure. There many technologies which have revolutionized the modern world of aerospace industries. However, the production of aerospace structures is mainly dependent on two leading technologies, i.e. the progress in the computational power of modern operating systems, and the developments in advanced composites manufacturing technology.

## COMPOSITE MATERIAL SYSTEMS FOR AEROSPACE INDUSTRY

The classification of composite material systems (reinforcements and matrices) for the aerospace industry is given in Table 7 and Table 8, respectively.

**Table 7:** Characteristics and applications of various space-grade reinforcing fibers

| Fibre | Density (glcc) | Modulus (GPa) | Strength (GPa) | Application areas |
|---|---|---|---|---|
| **Glass** | | | | |
| E-glass | 2.55 | 65-75 | 2.2-2.6 | Small passenger a/c parts, air-craft interiors, secondary parts; Radomes: rocket motor casings |
| S-glass | 2.47 | 85-95 | 4.4-4.8 | Highly loaded parts in small passenger a/c |
| **Aramid** | | | | |
| Low modulus | 1.44 | 80-85 | 2.7-2.8 | Fairings; non-load bearing parts |
| Intermediate modulus | 1.44 | 120-128 | 2.7-2.8 | Radomes, some structural parts; rocket motor casings |
| High modulus | 1.48 | 160-170 | 2.3-2.4 | Highly loaded parts |
| Carbon | | | | |
| Standard modulus (high strength) | 1.77-1.80 | 220-240 | 3.0-3.5 | Widely used for almost all types of parts in a/c, statellites, antenna dishes, missiles, etc. |
| Intermediate modulus | 1.77-1.81 | 270-300 | 5.4-5.7 | Primary structural parts in high performance fighters |
| High modulus | 1.77-1.80 | 390-450 | 2.8-3.0<br>4.0-4.5 | Space structures, control surfaces in a/c |
| Ultra-high strength | 1.80-1.82 | 290-310 | 7.0-7.5 | Primary structural parts in high performance fighters, spacecraft |

A variety of resin systems is being utilized in aerospace industry. Some polymeric matrices are given in Table 3.

**Table 8:** Polymeric matrices commonly used in aerospace sector

| Thermosets | | | | Thermoplastics |
|---|---|---|---|---|
| Forms cross-linked networks in polymerization curing by heating | | | | No chemical change |
| Epoxies | Phenolics | Polyester | Polyimides | PPS, PEEK |
| • Most popular<br>• 80% of total composite usage<br>• Moderately high temp.<br>• Comparatively expensive | • Cheaper<br>• Lower viscosity<br>• Easy to use<br>• High temp usage<br>• Difficult to get good quality composites | • Cheap<br>• Easy to use<br>• Popular for general applications at room temp | • High temp application 300°C<br>• Difficult to process<br>• Brittle | • Good damage tolerance<br>• Difficult to process as high temp 300-400° C is required |
| • Low shrinkage (2-3%)<br>• No release of volatile during curing | • More shrinkage<br>• Release of volatile during curing | • High shrinkage (7-8%) | | |
| • Can be polymerized in several ways giving varities of structures. morphology and wide range of properties | • Inherent stability for thermal oxidation<br>• Good fire and flame retardance<br>• Brittle than epoxies | • Good chemical resistance<br>• Wide range of properties but lower than epoxies<br>• Brittle<br>• Low $T_g$ | | |
| • Good storage stability to make prepregs | • Less storage stability-difficult to prepreg | • Difficult to prepreg | | • Infinite storage life. But difficult to prepreg |
| • Absolute moisture (5-6) causing swelling and degradation of high temp properties<br>• Also ultra violet degradation in long term | • Absorbs mositure but no significant effect of moisture in working service range | • Less sensitive to moisture than epoxies | | • No moisture absorption |

Pre-pregs are often utilized for the manufacturing of aerospace components through autoclave molding process. Filament winding is a popular technique which is used to manufacture composite cylinders and shells for unique applications such as rocket motor casing for missiles and space launch vehicles. Low-speed aircraft components are constructed from glass fiber composites which are manufactured through room temperature curing and oven curing processes. Composite tooling is abundantly used to produce a small number of components whereas metallic or ceramic tooling are preferred for bulk production. Radomes (radar domes) are usually manufactured from resin injection/transfer molding (Tretyakov & Sochava, 1993). Some of the modern systems are given in Table 4 along with the types of components where they are used in a typical high-performance aircraft. Standard composite systems used in aerospace sector are illustrated in Table 9.

**Table 9:** Composite material systems for aerospace applications

| Materials system | Application area |
|---|---|
| • 170°C curing high strength-carbon-epoxy<br>– Zero-bleed (neat resin content) UD prepregs<br>– 5 HS or 8 HS bi-directional fabric prepreg<br>– toughness, good out-life and shelf-life | Structural components of fighter aircraft and helicopters. e.g. wing skins, spars, fin, rudder, elevons, doors, etc. |
| • 175°C curing intermediate modulus carbon with epoxy + BMI/ cynate-ester<br>– Zero-bleed (neat resin content) UD prepregs<br>– 5 HS or 8 HS bi-directional fabric prepreg<br>– high toughness, good out-life and shelf-life<br>– low environmental degradation | Frames, stiffeners, rotor blades |
| • 120°Curing HS-carbon-epoxy<br>– Zero-bleed (neat resin content) UD prepregs<br>– 5 HS or 8 HS bi-directional fabric prepreg<br>– toughness, good out-life and shelf-life | Structural components of helicopters or transport aircraft1 e.g. spars, fin, rudder, elevons, doors, etc. Frames, stiffeners |
| • Aramid fibre in low polyester / cnate esters | Radome |
| • Cu-mesh epoxy prepreg | For Lightning Strike protection Wing-skin, others |

Composite materials exhibit superior strength-to-weight ratio (~up to 20% better) as compared to metals. The lower structural weights result in minor fuel consumption and emissions, ultimately augmenting the efficiency of the aerospace structures. Moreover, the composite structures are cheaper than metallic structures due to lower production costs and lesser joining operations. Composite materials attracted the aviation industry due to their tremendous breakthroughs in the construction sector. Initially, the use of composites was materialized by aircraft manufacturers for manufacturing high performance and maneuverable products (Edwards, 2008).

Weight plays a major role in aerospace applications, and constant struggles are being made to improve the lift to weight ratio of the aerospace structures. Composites materials have played an important part in the weight reduction of the structures, and the critical composite materials for aerospace applications are listed in Table 7, Table 7 and Table 9. There are some other categories of composite materials which can provide excellent structural characteristics for space environments. For instance, boron fiber composites are high strength hybrid composites which contain tungsten core inside the boron fibers. The application of composite parts in place of metallic components in commercial and private aircraft is growing rapidly (Kainer, 2006).

## USE OF COMPOSITES MATERIALS IN AIRCRAFT MANUFACTURING

Advanced composites play a vital role in the revolution of the aerospace industry. For decades, Airbus has cultivated the use of advanced composites and other innovative materials in aircraft manufacturing and design. The use of advanced structural materials has enabled Airbus to lead the aviation industry by developing economical and eco-friendly aircraft ranging from single-aisle Airbus-A320 to the 21st-century flagship, i.e. Airbus-A380.

The need for the usage of suitable materials for particular applications has urged the scientists and manufacturers to develop the innovative class of aerospace materials. It was reported that six types of materials were utilized for the manufacturing of Airbus A320 and A330. However, 20 different material systems were used for the production of Airbus-A380, which contains approximately 61% aluminum (lowest aluminum content among all Airbus models).

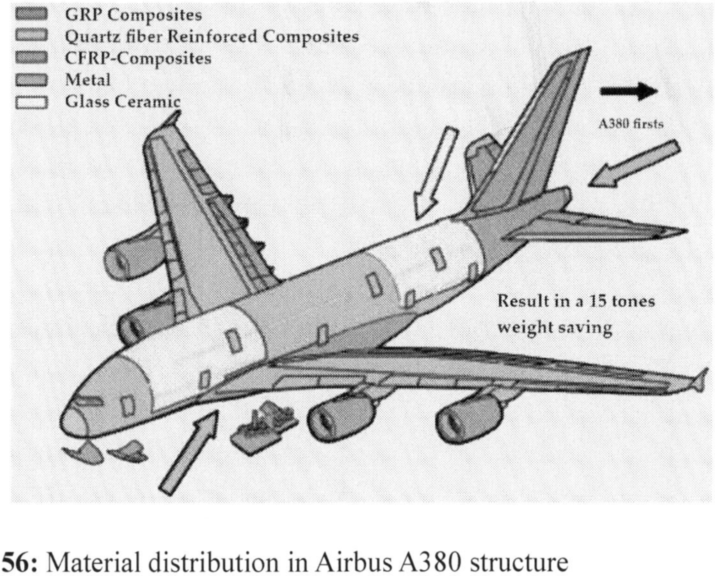


**Figure 56:** Material distribution in Airbus A380 structure

## Composite Technology and its Innovations in Airbus Jet-liners

Airbus is considered to be the pioneer of technological innovations in aircraft manufacturing and design. Airbus was the first company to use composite materials for constructing commercial aircraft such as Airbus-A310. Additionally, Airbus has administered a bulk scale usage of composite materials over the last three decades. For the comparison purpose, it was reported that Cornerstone utilized around 5% composite materials in the A300A310's aircraft during their initial production runs. On the contrary, Airbus has considerably increased the use of composite materials in Airbus-A380 (~25%) and Airbus-A350XWB (~53%).

## Composites Usage in the Airbus-A350XWB Structure

The structure of next generation Airbus-A350XWB contains 19% AL/AL-Li, 14% Titanium, 6% steel, and 53% composites. Carbon Fiber Reinforced Plastic (CFRP)-composites based skin panels are placed over the aluminum airframes while keeping the cross sectional area void. Aluminum is an excellent material to resist lightning strikes and radiations; therefore, aluminum strips are used in airframe structures. Fin assembly and horizontal stabilizers of the rear fuselage are constructed from carbon fiber composites as shown in Fig. 57.

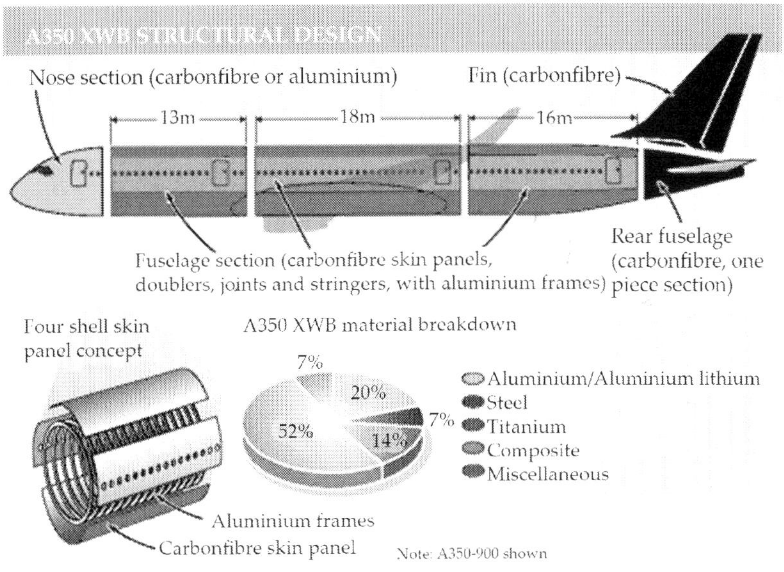


**Figure 57:** Configuration of composite structures in Airbus- A350XWB

CFRP composites have, broadly, been used in the manufacturing of doors, clips, fuselage panels and window frames while hybrid composite-titanium structure was used to manufacture door frame. Titanium was used for the first time in the manufacturing of the aircraft structure (Collings, 1982). Application of composite materials in Airbus-A350XWB has significantly increased the service life of aircraft from 6 to 12 years, with a substantial reduction in maintenance costs. Moreover, the incorporation of composites has considerably reduced the maintenance work related to fatigue and corrosion, which is significantly higher in conventional aluminum aircraft (Staszewski et al., 2009).

## Composite Usage in the Airbus-A400M Structure

Around 30% structure of the Airbus-A400M was manufactured from composites to reduce the overall weight of the aircraft. For the first time, a huge portion of wings was constructed from the composites. Composites are used to manufacture the entire tail structure including vertical and horizontal stabilizers and the control surfaces. The propeller blades, undercarriage bays, and the cargo doors are also produced from Kevlar fiber composites. Almost 19 m large aircraft's wings have also been manufactured from composite materials. Airbus-A400M has high payload capacity at lower weights due to the extensive use of lightweight composite structures (Mrazova, 2013).

## Composite structures in Airbus-A380

The key goal of every aircraft manufacturing company is to construct the lightest structures by utilizing the most suitable materials for specific applications. Composite materials are an excellent choice for producing high-performance, lightweight aircraft structures. The Airbus-A380 was the pioneer aircraft which consisted of a central wing box manufactured from Carbon Fiber Reinforced Plastic (CFRP) composites. About 1.5-ton reduction in weight is achieved when CFRP-composites are used in place of aluminum alloys.

The pressure bulkhead and deck floor beams are also manufactured from CFRP-composites. Resin Transfer Molding and Automated Fiber Placement were utilized for production of Airbus-A400M structures.

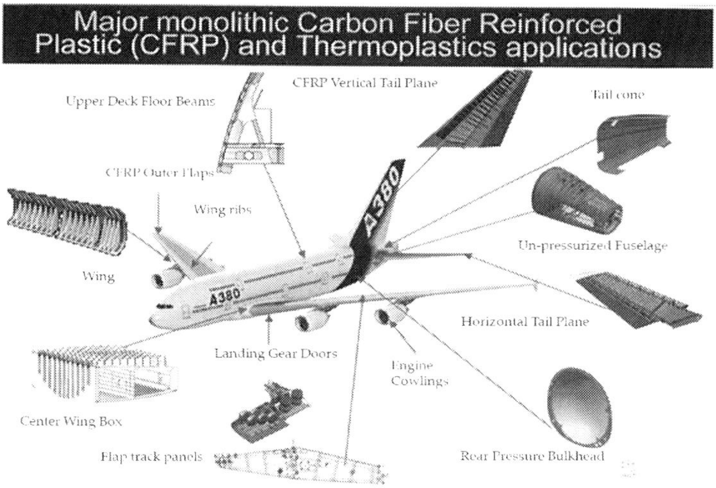


**Figure 58:** Major monolithic CFRP and Thermoplastics applications

## SPACE APPLICATIONS OF COMPOSITE MATERIALS

Satellite and launch vehicle structures are mainly manufactured from metallic alloys. Satellite components and boxes are made of aluminum alloys while launch vehicle casings are made from managing steel. However, composite materials have gained vital importance for providing lightweight structures. Therefore, most of the aerospace components are being manufactured from composite materials these days (Kodiyalam, 1996). Currently, graphite fiber composites, carbon fiber composites, and metal matrix composites are being utilized extensively for space applications. Some common space applications of composites are listed below:

- Space telescopes
- Space station
- Antenna systems
- Space tug
- Cryogenic tanks
- Small and lightweight satellites

## Applications of Composites in Satellite Structures

The primary use of graphite fiber reinforced aluminum (Gr/Al) composites in satellites include telescope antenna boom. The boom structure of famous Hubble Telescope contained diffusion bonded graphite fibers in aluminum 6061 matrices. A satellite image of Hubble telescope is shown in Fig.59.

**Figure 59:** Satellite image of Hubble Telescope

Composite boom structure (306 m long) provides desired strength, stiffness, and coefficient of thermal expansion to facilitate controlled positioning of the antenna during space trips. Metal matrix composites (MMCs) provide excellent waveguide function by enabling smooth transmission of electrical signals between the antenna dish and spacecraft. Additionally, MMCs provide thermal, mechanical and dimensional stability to the structure with a dimensional tolerance of around ± 0.15 throughout the entire length of the structure (Lukez, 1987).

Similar to Gr/Al composite boom structure, some novel MMCs have been designed and manufactured to serve multidimensional requirements for thermal, structural, electrical applications. For instance, a prototype of radiators was developed from Gr/Al composites for thermal, structural and EMI shielding applications (Leger et al., 1987). Moreover, thermally conductive Gr/Cu composites have been developed to provide high thermal conductivity in structural radiators (Lukez et al., 1987).

Discontinuously reinforced aluminum (DRA) composites contain two reinforcements in a single matrix. These type of composites were initially manufactured from SiC particulates and whiskers of SiC in an aluminum matrix during the 1980s. Potential uses of discontinuously reinforced aluminum (DRA) composites include attachment and joint fittings for truss structure, electronic packages, longerons, thermal planes, bushings and mechanism housings. Truss structures are typically produced from near net shape technique. DRA composites are specially designed for thermal management and electronic packaging applications due to their high thermal conductivity, a tailorable coefficient of thermal expansion and low density (Thaw et al., 1987; Miracle, 2001). Presently, many Gr/Al and SiC/Al electronic packages are being manufactured and installed in Global Positioning System (GPS) satellites and communication satellites. These unique composite components are lighter and cost efficient as compared to conventional metallic components.

DRA composites are also used for regulation of thermal power in semiconductor modules of geosynchronous earth orbit (GEO) communication satellites. These composites have replaced Cu/W alloys due to their low weights (up to 80% weight saving) and maneuverable thermal conductivities. Keeping in view the unique characteristic of advanced composites, the applications of MMCs and DRA composites, in satellite components, are expected to grow.

## Use of Composite Materials in Launch Vehicles

Space launch vehicles (SLVs) have a unique architecture which requires structural and thermal protection combined with high mechanical strength, excellent fatigue resistance, lightweight and minimum possible manufacturing cost. Innovations in launch vehicle systems are needed to address particular challenges including a reduction in structural mass, management of structural margins, retention of cryogenic oxygen and hydrogen fuels and development of robust thermal protection systems. Global space agencies are focused on the development of next-generation launch vehicles with lighter upper stage structures. Advanced research is being carried out to investigate the novel class of materials and processing techniques for provision of higher structural, and thermal efficiencies in SLVs (Mazzola, 2012). Some advanced technologies are being explored to manufacture high-performance vehicles with reduced weight. Typical SLV configuration is shown in Fig. 60.

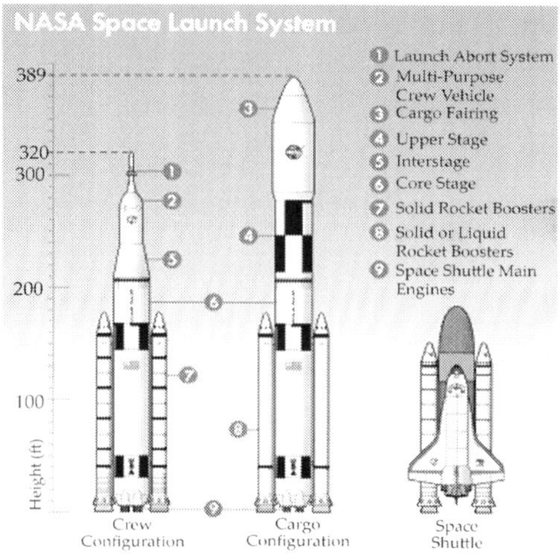


**Figure 60:** Typical configuration of launch vehicle and space shuttle

The latest reusable space launch vehicles utilize lightweight and large structures developed from metals and composites. Currently, various CFRP-based SLVs structures are under investigation such motor casings, cryogenic tanks, inter-stages, casing rings and thrust frames.

The performance of a launch vehicle directly depends on the inert weight of its upper stage. It is essential to use lightweight upper stage structure to maximize the effective payload mass. CFRP composites provide excellent specific stiffness and strength as compared to conventional metals. Therefore, they are recognized as the most suitable materials for manufacturing of upper stage cryogenic tanks. The lower thermal conductivity of the composites makes them an ideal choice for development of lightweight under-fairing stage which is designed to resist high-temperature exposures. Development and maturation of advanced composite technology actively rely on processes improvement, material assessment, and structural design innovations.

Besides the investigation of lightweight wall concepts (sandwich, multiwall, skin), several issues need to be addressed before the development of cryogenic tanks from composite materials. Some major concerns, regarding the development of launch vehicles, are illustrated below:

- The permeability of liquid hydrogen through the tank wall is a major concern. The issue can be resolved by introducing a tight and thin liner compatible with cryogenic liquids.

- Wear resistance of interfaces and singular points of primary structures
- Close assembling of cryogenic tank parts
- Integration of hot and cold thermal protection
- Health and damage monitoring
- Operability and reusability

Presently, applications of CFRPs for the development of liquid oxygen (LOX) containers are being explored. Compatibility of LOX with CFRPs and other composite systems is also being researched. Various CFRP structures for SLV applications have already been manufactured and validated by evaluation tests (Robinson, 1994).

**Figure 61:** CFRP fuel tank developed by SpaceX

At present, inter-tanks are also being constructed from composite materials (Hou & Gramoll, 1999). The key features of such full-scale composite structure are illustrated below:

- Inter-stage is a large symsmetrical or non-symmetrical structure, with a dimensional range between 5 to 8 meters.
- It is a bulk composite structure capable of withstanding thermal and mechanical loads encountered due to inertial effects of cryogenic tanks and dynamic loading during space maneuvers.
- The introduction of local loading conditions in inter-tanks

requires them to be fatigue resistant.

- Inter-tanks experience thermal gradients due to the presence of thermal protection systems (TPS) and cryogenic tanks.
- Manufacturing of inter-stage structures is typically carried out by automated filament placement (AFP) technique.

Due to the increasing demand for lightweight aerospace structure, the development of thrust frames from of full-scale composite materials is essential nowadays. The main features thrust frames are discussed below:

- Thrust frames (chambers) are heavily weighted structural components in a launch vehicle system experiencing various design and structural loads, i.e. unsymmetrical cut-off design, and thrust induced mechanical loads.
- Harsh environmental conditions encounter thrust Chambers due to their close locality with a thermal, chemical, acoustic and vibrational interference of the engine.
- Thrust chamber is a complex structure integrated from a large variety of interfaces.

Thrust frames are typically manufactured from CFRPs (carbon fiber reinforced in Bismaleimide or BMI resin). However, some metallic and MMC parts are also integrated into thrust frame structures CFRPs possess, i.e. good fracture resistance, excellent stiffness, intermediate toughness, and ability to operate at high temperatures without. Thrust frames are usually manufactured from carbon fiber composites due to a range of properties which any degradation.

**Figure 62:** Thrust frame constructed from CFRP composites

Another SLV component is wing box, which experiences high-temperature exposure and high loading conditions during its service. Before the development of composite materials, it was a challenge to design and manufacture a high-performance wing box. Presently, wing boxes are usually produced from carbon fiber composites (CFRPs with BMI matrix) due to their excellent stiffness and strength.

Composites are most appropriate materials for manufacturing of reusable structures to serve as thermal protection systems (TPS). There are various reinforcement/matrix combinations which can be opted for development of TPS composites. Typically ceramic matrix composites are given precedence over other categories of composites due to their high thermal resistance. Carbon fiber or silicon carbide (SiC) are usually reinforced with a ceramic matrix to produce a rigid and thermally resistant structure. Extensive research is underway to develop and evaluate TPS composites from various reinforcement/matrix combinations. It has been reported that CMCs based TPS systems have been successfully manufactured and tested for current SLV applications (Blosser, 1997). Moreover, repair techniques for TPS structures are also being developed.

## CHALLENGES FOR COMPOSITE APPLICATIONS IN AEROSPACE SECTOR

Problems regarding the expanded use of composites in aerospace applications are shared by space agencies, aviation industries and defense sectors (Harris et al., 2002). Some of the major concerns regarding aerospace composites are discussed below:

- Manufacturing and equipment costs are key factors in limiting the growth of the composite industry.
- Composites are extensively used for research and development work in defense sectors as compared to their commercial usage in aerospace industry.
- Empirical designs, maintenance technology, and standardization procedures are also critical factors for production of aerospace components.
- The lack of hierarchical design techniques and material standardization may result in the development of overdesigned aerospace products, which are comparatively less economical

than conventional metallic products.

- Certified but repetitive failure analyses are performed to validate the fatigue life of the composite structures.
- Composite applications are also constrained due to a limited human resource with composite handling skills and experience.

## EFFORTS TO COPE WITH MODERN CHALLENGES IN AEROSPACE COMPOSITES

It is essential to acknowledge the recent developments in the composite industry which have provided viable solutions to combat the current problems. Recently, NASA has conducted a successful test of a composite aircraft wing at Langley research center. The structure of the wing was designed to fulfill the requirements of a commercial aircraft with a capacity of 220 passengers (Tenney et al., 2009). Graphite fiber preforms and Kevlar fibers were used to manufacture a 41 feet long wing box via resin-film-infusion at maximum resin impregnation rates. Test results showed that about 97% of the ultimate design load was sustained by the wing structure, which was considered acceptable; and further design modifications were made accordingly. Boeing and Airbus adapted the innovations in composite manufacturing to develop cost competitive and efficient aircraft structures.

It is imperative to note that NASA spent its enormous resources to evaluate the performance of the composite wing structure. However, the research proved expedient for future aerospace endeavors due to the standardization of materials, processes, and tests. The shortcomings in composites technologies can be overcome by developing more cost efficient and performance specific equipment and methods.

## TECHNOLOGY INTEGRATION

The future of aerospace industry heavily depends on the integration and exploitation of various advanced techniques (Tenney & Hernandez, 1998). Range, survivability, and agility of an aircraft can be improved significantly by utilizing modern materials and technologies such as smart composites, flexible structures, forebody vortex control, synthetic jets, passive porosity, advanced control laws, fluidic thrust vectoring and continuous Mold-line technology. For commercial aircraft structures, these technologies offer active shape control, buffet load alleviation, health monitoring, thrust vectoring, active transition control, inlet and nozzle shaping, vibration suppression, exterior noise suppression, active separation control, flutter suppression and gust load alleviation.

Integration and application of the advanced technologies for manufacturing of high capacity transport aircraft will result in a superior performance as compared to conventional aircraft. Excrescence drag reduction, design optimization, and laminar flow control will result in 4.6% reduction in the gross takeoff weight (TOGW) for commercial aircraft (McKinley, 1999). Development of composite tails and wings, light weight landing gear, composite fuselage and advanced alloys for radiation shielding will shrink the takeoff gross weight by 24.3%. Innovations in computational fluid mechanics, propulsion system, material characterization and aero-mechanical design can materialize a weight saving of 13.1%. Additionally, the advancements in the static stability, power usage, control systems, navigation systems, and intelligent flight systems would result in 9% reduction of the TOGW. These advanced technologies collectively account for 51% weight reductions in aircraft and spacecraft structures. On the contrary, only composite technology coupled with structures technology can result in 48% weight reduction.

Incorporation of advanced composite and structures technology can give rise to 45.7% weight (TOGW) reduction for a high capacity/long haul subsonic aircraft with 8500 nautical miles range, a passenger capacity of 800, and 10,000 feet landing requirement. Laminar flow control, excrescence drag reduction, and design optimization will yield 11.8% weight savings. Composite fuselage, composite wings, lightweight landing gears and aeroelastic tailoring will result in the reduction of TOGW by 19.1%. %. Advances in computational fluid mechanics, material characterization, propulsion systems and aero-mechanical design can materialize a weight saving of 12.2%. The advancements in the static stability, power usage, control systems, navigation systems, and intelligent flight systems would result in 2.6% reduction of the TOGW. It is quite evident from the statistics that a blend of composites and structure technologies is essential for developing a high-performance aerospace structure (Noor et al., 2000).

## THE FUTURE OF COMPOSITE MATERIALS IN AEROSPACE INDUSTRY

Environmental lobbying and escalating fuel costs are driving factors for commercial aircraft industries to manufacture lightweight sustainable structures to achieve high-performance goals. Besides operational costs of airlines, aircraft maintenance costs can be reduced by using corrosion and wear resistant composite components. The global competition in aviation and aerospace sectors has forced the manufacturing industries to explore the novel technologies for developing cost-competitive products. Innovative materials are being developed to cope with the modern challenges of the aerospace industry. New materials are classified as the materials which are being applied in other sectors, but they are yet to be employed in 'as-designed' aerospace applications. Some of the new materials including metal matrix composites (MMCs) and ceramic matrix composites (CMCs) are being tested for aerospace use, but their wide-range recognition is still pending due to various reasons. Some potential materials for aerospace applications are discussed below:

### Ceramic Matrix Composites (CMCs)

CMCs are suitable materials for high-temperature applications. CMCs exhibit excellent thermal properties with enhanced mechanical properties. Limitations of conventional ceramics (i.e. low toughness) are overcome by using fiber reinforced ceramics composites. Potential applications of CMCs in aerospace and aviation industry include turbine disks, turbine aerofoils and combustor liners (Peel, 1992).

### Metal Matrix Composites (MMCs)

MMCs are typically comprised of metal matrices reinforced with ceramic fibers. The most common metal matrix materials are aluminum and titanium while the reinforcements are usually the oxides, nitrides or carbides of metals. Metal matrix reinforced with carbide, nitride or oxide fibers has many benefits over solid metals (Naslain et al., 2004). But they are comparatively hard to machine due to their reduced toughness. Potential applications of MMCs include load bearing surfaces such as turbine fan blades, rotor blades, and floor supports in aircraft and spaceships.

## Carbon Nanotubes (CNTs)

Carbon Nanotubes (CNTs) are deemed as next generation advanced materials. CNTs have been used extensively in electronic and semiconducting applications such as solar cells, sensors, etc. However, CNTs have been used scarcely in space applications due to lack of guidance and knowledge regarding their use. Wide research is underway to evaluate the applications of CNTs in electronics and EMI shielding of commercial aircraft (Meyyappan, 2000).

Besides their lightweight, aluminum structures are considered to be resistant to electromagnetic radiations. However, some radiations are trapped inside the aircraft or satellite structure which can damage the electronics. Therefore, it is essential to shield the structures both internally and externally. CNTs based composites are excellent materials for EMI shielding applications. Consequently, composite aerospace structures embedded with CNTs are lightweight and immune to space radiations.

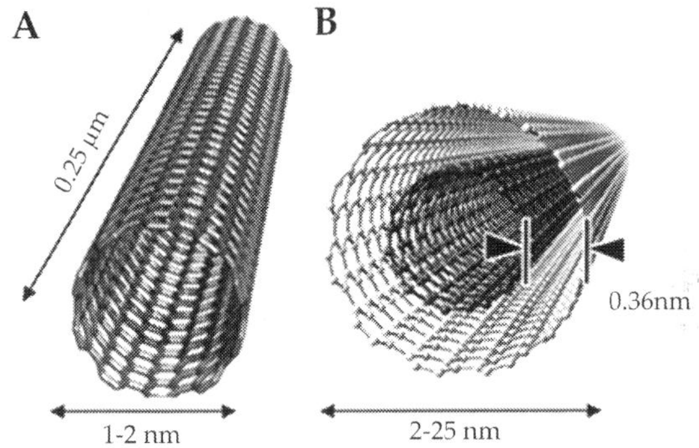


**Figure 63:** Structural configuration of single-walled and double-walled carbon nanotubes (CNTs)

Carbon fibers have higher strength and stiffness values which make them susceptible to brittleness at extreme impact loads. On the contrary, CNTs are as stiff as carbon fibers, but they have improved flexibility and toughness. Fabric form of CNTs can be used to develop composites for versatile aerospace applications. CNTS are multifunctional materials which can be utilized in aircraft, spacecraft, and satellites for various applications such as antennas, wing structures, and electronic devices.

## Shape Memory Alloys (SMAs) and Hybrid Composites

Shape memory alloys are a unique class of materials with extraordinary ability to revert to their original (pre-deformed) shape after application of plastic loads (Roh et al., 2004). SMAs typically consist of nickel/copper alloy systems, though other alloy systems can also be used. The combination of SMAs with CNTs can result in a composite material with excellent hybrid characteristics of both constituents. These hybrid composites can be used to develop actuators for launch vehicles and aircraft. Traditional materials cannot serve the purpose of hybrid composites because of weight limitations.

## Honeycomb Sandwich Structures

Aerospace sandwich structures consist of two face sheets adhesively bonded with a core structure, forming a composite structure. There is a range of the core structures which can be used for aerospace applications, but, the honeycomb structure is the most common core material. A typical honeycomb sandwich structure is shown in Fig. 64.

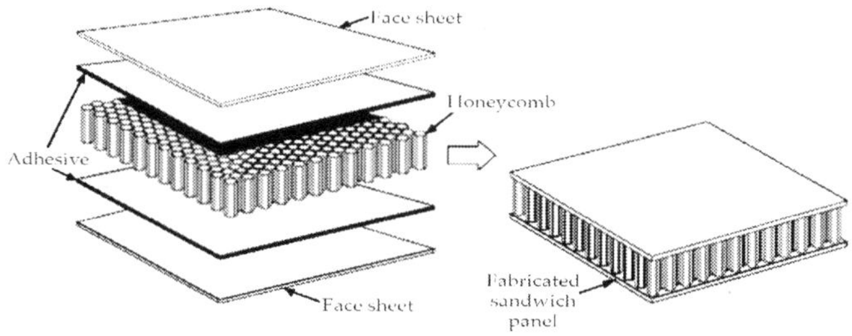


**Figure 64:** Typical honeycomb sandwich panel for satellite applications

Honeycomb structures are manufactured from both metallic and nonmetallic materials depending on the requirements of an aerospace component. Aluminum, Nomex, and fiberglass are the most common honeycomb materials. There are a variety of ways and configurations to manufacture honeycomb sandwich structures due to their flexible design characteristics. A large combination of face sheet materials and core materials can be bonded to make tailor-made composite structures (Price et al., 1997). Honeycomb sandwich panels provide increased stiffness at comparatively lower weights than conventional composites.

Similar to a beehive, the geometric structure of honeycomb sandwich panels impart strength and weight saving to the structure while providing support to the prepreg skin sheet. The unique structure of honeycomb provides exceptional compressive strength along the cell wall direction and excellent shear strength along the direction perpendicular to the cell walls. An adhesive film is usually used to bond honeycomb core with pre-preg sheets. An adhesive layer is also deposited on the surface of the sandwich panel to enhance the aesthetic appearance, uniformity, and smoothness of the surface (Hayes & Gammon, 2010).

## REFERENCES

1. Asundi, A., & Choi, A. Y. (1997). Fiber metal laminates: an advanced material for future aircraft. Journal of Materials Processing Technology, 63(1-3), 384-394.
2. Baker, A. A. B. (2004). Composite materials for aircraft structures. AIAA.
3. Blosser, M. L. (1997). Development of metallic thermal protection systems for the reusable launch vehicle. In AIP Conference Proceedings, 387(1), 1125-1144.
4. Collings, T. A. (1982). On the bearing strengths of CFRP laminates. Composites, 13(3), 241-252.
5. Edwards, T. (2008). Composite materials revolutionize aerospace engineering. Ingenia, 36, 24-28.
6. Harris, C. E., Starnes, J. H., & Shuart, M. J. (2002). Design and manufacturing of aerospace composite structures, state-of-the-art assessment. Journal of Aircraft, 39(4), 545-560.
7. Hayes, B. S., & Gammon, L. M. (2010). Optical microscopy of fiber-reinforced composites. ASM International.
8. Hou, A., & Gramoll, K. (1999). Design and fabrication of CFRP interstage attach a fitting for launch vehicles. Journal of Aerospace Engineering, 12(3), 83-91.
9. Kainer, K. U. (Ed.). (2006). Metal matrix composites: custom-made materials for automotive and aerospace engineering. John Wiley & Sons.
10. Kodiyalam, S., Nagendra, S., & DeStefano, J. (1996). Composite

sandwich structure optimization with application to satellite components. AIAA Journal, 34(3), 614-621.

11. Leger, L., Visentine, J., and Santos-Mason, B. (1987). Selected Materials Issues Associated With
12. Space Station. SAMPE Quarterly, 18(2), 48-54.
13. Lukez, R. (1987). The Use of Graphite/Epoxy Composite Structures in Space Applications.
14. Lukez, R., Nelson, D., Teller, V, and Rockoff, H. (1987). Design, Fabrication, and Testing of Rolled Carbon/Epoxy Struts for Space Station Applications. 19th International SAMPE Technical Conference, Washington, D.C.
15. Mayer, N. J. (1974). Carbon composites in space vehicle structures.
16. Mazzola, M., Biagioni, M., & Mazzola, L. (2012). Composite Materials in Launch Vehicles. Wiley Encyclopedia of Composites.
17. McKinley, R. E. (1999). Blended-wing-body- Design challenges for the 21 st century. In International Conference on Nonlinear Problems in Aviation and Aerospace, 2nd, Embry-Riddle Aeronautical Univ, Daytona Beach, FL (pp. 473-478).
18. Meyyappan, M. (2000). Carbon Nanotubes for Space Applications.
19. Miracle, D. B. (2001). Metal matrix composites for space systems: current uses and future opportunities. Affordable Metal-Matrix Composites for High-Performance Applications II, 1-22.
20. Mrazova, M. (2013). Advanced composite materials of the future in the aerospace industry. Incas bulletin, 5(3), 139.
21. Naslain, R., Guette, A., Rebillat, F., Pailler, R., Langlais, F., & Bourrat, X. (2004). Boron-bearing species in ceramic matrix composites for long-term aerospace applications. Journal of Solid state chemistry, 177(2), 449-456.
22. Noor, A. K., Venneri, S. L., Paul, D. B., & Hopkins, M. A. (2000). Structures technology for future aerospace systems. Computers & Structures, 74(5), 507-519.
23. Peel, C. J. (1992). Metal Matrix Composite for Aerospace Applications. The Institute of Materials (UK), 41-51.
24. Perret, B., Schartel, B., Stöß, K., Ciesielski, M., Diederichs, J., Döring, M., & Altstädt, V. (2011). Novel DOPO-based flame

retardants in high-performance carbon fiber epoxy composites for aviation. European Polymer Journal, 47(5), 1081-1089.

25. Pipes, R. B. (2002). Cyber Technology for Materials and Structures in Aeronautics and Aerospace.
26. Price, T. L., Dalley, G., McCullough, P. C., & Choquette, L. (1997). Handbook: Manufacturing Advanced Composite Components for Airframes (No. DOT/FAA/AR-96/75). Federal Aviation Administration Washington Dc Office of Aviation Research.
27. Robinson, M. (1994). Composite cryogenic propellant tank development. In 35th Structures, Structural Dynamics, and Materials Conference (p. 1375).
28. Roh, J. H., Oh, I. K., Yang, S. M., Han, J. H., & Lee, I. (2004). Thermal post-buckling analysis of shape memory alloy hybrid composite shell panels. Smart Materials and Structures, 13(6), 1337.
29. Soutis, C. (2005). Fibre reinforced composites in aircraft construction. Progress in Aerospace Sciences, 41(2), 143-151.
30. Staszewski, W. J., Mahzan, S., & Traynor, R. (2009). Health monitoring of aerospace composite structures–Active and passive approach. Composites Science and Technology, 69(11), 1678-1685.
31. Tenney, D. R., & Hernandez, G. (1998). 21 st Congress of the International Council of Aeronautical Sciences. Melbourne, Australia.
32. Tenney, D. R., Davis Jr, J. G., Pipes, R. B., & Johnston, N. (2009). NASA composite materials development: lessons learned and future challenges.
33. Tretyakov, S. A., & Sochava, A. A. (1993). Proposed composite material for nonreflecting shields and antenna radomes. Electronics Letters, 29(12), 1048-1049.
34. Thaw, C., Minet, R., Zemany, J., & Zweben, C. (1987). Metal Matrix Composite Microwave Packaging Components. SAMPE JOURNAL, 23(6), 40-43.

# CHAPTER 7

# MILITARY AND DEFENSE APPLICATIONS OF COMPOSITE MATERIALS

## INTRODUCTION

Worldwide soldiers are sworn to perform their duty of guarding their countries. Many manufacturers are working on the development of high standard composite components to shield the soldiers. Composite manufacturers have developed new and modern methods of blending materials to fabricate vehicles as per military standards and requirements (Hogg, 2006). For shielding of vehicles and military equipment, high-quality steel has been long in use. However, the use of steel is not very desirable due to multiple reasons. The heavy weight of steel gives rise to the logistical problems in moving the vehicles to the war site. The approximate weight of a typical army vehicle is around 60t and thus can limit the performance of such vehicles; moreover, as the use of smart weapons and threats as a result of it have increased over the years, it is now important more than ever, to be able to get in and out of the vehicle in case of an emergency or in war field. Therefore, increased weight makes the survival of vehicles difficult in the harsh environment of the field. Hence, it is important to modify the suspension system in order to balance the amplified inertial energy. Through cooperation between composite manufacturers and customers, clients can be involved in the complete development cycle of the desired product from design to testing, further to analysis and then to ultimate product. In this way, composite manufacturers can deliver the end product with same features as desired by the customer for any particular application.

The above mentioned problems have led to a significant increase in the manufacturing of composite armored vehicles. Composite coatings like PPFR and PUFR are used to provide an additional protection layer to a wide range of artillery, vehicles, boats, and structures (Cohen & Israeli, 1998). In addition to the polymer matrix, several other composite materials including ceramic matrix, metal matrix, and carbon composites are in use for manufacturing of military equipment and vehicles.

For example, carbon composites are used in the manufacturing of space vehicles and aid them to withstand heat upon their reentry into space. It also has a very significant applicability in Army weapon systems where overheating causes the failure of systems. In addition to the major advantage of composite materials in terms of their heat withstanding capability, an added benefit is that they are much light in weight than steel components. Main advantages of using composite materials are:

- Reduction in weight
- Fuel efficiency
- High payload
- High performance and speed proficiency

Cost and weight are the key parameters of the military vehicles and their components. Another important parameter is the strength. While constructing military vehicles, manufacturers keep the protection factor of such vehicles in mind and in doing so, vehicles are made bulky. The use of composites, on the other hand, makes them light. Composites are resistant to corrosion and do not rust which helps their survival in harsh environments (Szczygieł & Kołodziej, 2005). Additionally, they have anisotropic strength properties which make them stronger in one direction. Hence, composite fibers can tolerate multi-directional loads and forces when multiple layers are stacked in different directions in them.

## RESEARCH IN COMPOSITE MATERIALS FOR MILITARY APPLICATIONS

Main materials which are employed in the construction of composite structures are:

- Kevlar
- Fiberglass
- Carbon fiber with different orientations.

A commercially used composite material is a low-cost composite comprising of glass fiber and PET resin. The radio transmit ability of this composite material allows it not to obstruct the other radio waves. The weight of glass fiber is almost 20% to 30% less as compared to steel. In order to reduce the weight up to 50% to 60%, carbon fiber composites must be used. A worthy challenger of titanium is a carbon fiber composite (PEEK matrix), which is used in high-end military vehicles (Molitor et al., 2001). On the other hand, carbon fiber nylon has been proved an equable competitor of aluminum.

In the field of ballistic operations and their protection, Norplex-Micarta is used which provides high-performance thermoset sheets of composite materials and their molded contours to fulfill the growing requirements of ballistic protection. Further, two other ballistic resistant composites, named, NP522 and NP523 are manufactured from fiber glass in a way that they comply with the strict military requirements.

The Norplex-Micarta Shot Blocker materials are presented as entirely-treated, stiff sheets, or as pre-pregs. Thicknesses of sheets range from 0.062 to 8.0 inches, whereas sheet sizes range from 48 to 120 inches. The trend of putting a sizeable amount of money to research and develop composite materials is increasing in the United States, both in government and the private sector. Few examples are mentioned below: A research grant was awarded to The University of Delaware to develop innovative and modern composite technologies in order to build up refined watercrafts that will be faster, stronger, cheap, stealthier, field repairable and have long service life. A composite watercraft is displayed in Fig.65.

**Figure 65:** Lightweight and high-speed composite watercraft

INVISTA S.a.r.l. was awarded a research grant for improving the already prevalent protection and safety measures of soldiers by manufacturing uniforms that are more durable, flame resistant, and are economical and thus provides protection against extemporaneous explosives devices. Another awardee of the research funding is Armor Dynamics who upon receiving the research funding are to develop advanced composite and reactive armor. Tex Tech Industries got funding to manufacture the core technology of ballistics which resulted in increased protection of troops against their destruction from bursts of bullets or blasts. Another government organization known as Hodgdon Defense was awarded research funding to research and then develop the lightweight composites which can be used to reduce the weight of high-speed military aircrafts.

## BENEFITS IN EXTREME SERVICE CONDITIONS

As mentioned above, reduced weight of composites makes their use very desirable in the manufacturing of mobile equipment, permitting increased efficiency of fuel consumption and payload, etc. When the weight of a vehicle is reduced, it improves its acceleration and deceleration which also greatly benefits the suspension system of such vehicle (Mangino et al., 2007).

Fatigue life of steel is approximately 1 billion cycles, and that of aluminum is estimated to be around 1 million cycles, and therefore, aluminum can be used as a replacement of steel. Some components, for example, the driving shaft of a car, undergoes a lot of cycles in a short span of time. Fatigue life of composite materials can be accurately said 'infinite' and therefore ideal to use in extreme conditions. Two other primary parameters which are critical in designing the composites are:

- Stiffness-to-weight-ratio
- Strength-to-weight-ratio.

Advantages of composites over metals are well known in both of these categories. To increase stiffness to weight or strength to weight ratio in a composite, fibers can be oriented along a path which is either dominant in terms of strength or stiffness. Most people are unaware of the fact that in most designed structures, for example, door or frame of a vehicle, the key design parameter is the stiffness of the panel and not its strength. Therefore, it is safe to say that stiffness to weight ratio is the cardinal design variable that an engineer needs to optimize to increase the performance (Jones, 1998).

## CUSTOMIZED PROPERTIES

Most of the composite materials achieve their efficacy in terms of low cost and weight properties by orienting their fibers in the direction of primary load, whereas, only sufficient fibers are oriented in the other directions to hold the inferior loads and keep the structure collected. According to Brad McPhee, president, and CEO of Creative Composites, "Strength is a given for military applications." The reason that combat vehicles are heavier than other vehicles is that they are little overdesigned to provide additional protection against the harsh environment of the field. Therefore, composites are used to reduce the overall weight of the vehicle. Contrary to the most commercial applications, the military is ready to pay substantial premiums for reducing weight due to the critical nature of their work.

There can be a countless number of materials which can be employed in composites. The creative range of composite materials varies from Kevlar to fiberglass to carbon fiber. Carbon fiber nylon is an average performance composite material used by military and costs around $20 per pound of material. For applications which need high-performance components, carbon fiber PEEK (Poly-ether-ether-ketone) is used. This material has high toughness properties but is comparatively expensive, costing about $100 per pound of material. This material also serves as an excellent competitor for titanium. An equable competitor of aluminum is glass fiber PET (Yang et al., 2011). To say that it is better to use than aluminum would not be an overstatement. However, it depends on the application where these materials are supposed to be used.

In 2007, Armor Works pronounced the manufacturing of low cost, and high standard composite armor suite to be used by the U.S Marine Corps in their new Expeditionary Fighting Vehicle abbreviated as EFV. The ballistic efficacy of employed hybrid composite armor used in the manufacturing of EFV offers distinctive amalgamations of unlike composites that provide lightweight protection. Backings which are used in ceramic and fragment armor systems decreases their cost and increases their multi-hit protection and structural veracity of these armors against armor piercing bullets, extemporary explosive devices, and military fragments.

CMCs based military armor vehicle is shown in Fig. 66.

**Figure 66:** Military armor vehicle constructed from ceramic matrix composites

The ceramic composite panels of MKU's are manufactured from ceramic tiles arranged below multiple layers of material. The advanced ceramic composite (ACC) panels used in it are made up of composite materials that are light in weight, have high performance, higher multi-hit abilities, and capable of performing well in wide temperature range. On the other hand, the standard composite panels of MKU's delivers the same protection and performance level at a relatively reduced cost, by compromising the lightweight benefits of ACC panels. Manufacturing of ballistic sheets initializes by feeding woven fiberglass into a pool of phenolic resin which is afterward passed through multiple rollers to get rid of excess resin. This fabric is then placed in a 150 ft. oven which burns off volatiles and leaves the material in a semi-cured form, and afterward cropped to the desired dimensions. These freshly manufactured sheets are then stacked, weighed and pressed to form a solid sheet which can be employed in different applications for constructing ballistics. Companies like Kanpur and Armor Works LLC manufactures military vehicles armors by using ceramic based composites which are light in weight. These materials are brittle in nature and provide extreme hardness and additional resistance to materials like high kinetic projectiles.

## COMPOSITE MATERIAL DEVELOPMENT

Two or more than two materials are fused together to develop composite materials. Most composites have matrix or reinforcement. The admirable properties of composites are due to reinforcements which are generally much stronger than the matrix. The purpose of the matrix is to hold the reinforcements in an organized manner.

Generally, any kind of plastics are used in the composition of the matrix; therefore such composites are often known as reinforced plastics. Although metals and ceramics have also been used as materials for matrix development, the most popular material is still plastic. The two most widely used plastic matrices are polyester resins and epoxy resins. Composites are presented in the form of plies or lamina. A single ply is composed of fibers which are oriented in a single (known as unidirectional), or in two directions (known as bidirectional).

Properties of composites behave best in the direction of the fibers whereas matrix properties dominate in cases orthogonal or transverse to the fibers. Since most of the times, structures are not loaded in a uni-direction; therefore, it is vital to place fibers in more than one directions. This is achieved by loading several plies together. A stack of this kind is known as a laminate. Efficient composites (low cost and light weight) mostly have their fibers directed in the orientation of primary load and just sufficient fibers in the other directions to transport secondary loads, therefore, to keep the whole structure together.

Nowadays several military defense applications depend on composite materials due to their distinctive properties. PRF Composite Materials is a famous composite manufacturing company which has been using composites for the development and manufacturing of military and defense products. Over the past 30 years, PRF has delivered materials for armored vehicles, military air crafts, and submarines to its clients leading high-security projects (Tucker, 2012). Image of fifth generation Russian submarine is shown in Fig.67.

**Figure 67:** Fifth generation Russian submarine developed from composite materials

## MANUFACTURING TECHNOLOGY

For many applications of composite materials, PRF has manufactured a variety prepreg resin systems like thermoplastic resins, phenolic and epoxy resins. It is no surprise that most of the military applications usually need specific considerations. The experience of manufacturing modified and tailored woven (Fig.68) and non-woven fabrics is vast and more work can be done in collaboration with clients to design and develop solutions ideal to the customer's requirements. Such modified materials can be manufactured using new and innovative hybrid woven making use of special fibers and materials. An additional advantage of using these materials is that they can also provide EMI and RFI shielding (Kritchevsky et al., 1987).

**Figure 68:** Woven Kevlar fabric used for the manufacturing of bullet proof jackets and helmets

PRF provides cutting services for kits made of composite materials to be used in ballistic grade fibers like Kevlar® and Dyneema ®. These materials can be cut down in either dry form or the form of prepreg; layers are added afterward. Maximum limit up to which we can cut ballistic grade Kevlar® is of 25mm in thickness. These kits can be supplied in bags with proper labeling and thus ready for use. Main benefits of kit cut composites are reduced price and manufacturing time, steady quality and optimal usage of material.

## Qiicote Mold Release Technology

To manufacture composite materials made of thermoset and thermoplastics, PRF has developed Qiicote mold release technologies which have aided in the development of advanced composite products. Qiicote products offer multiple advantages, some of which are:

- Qiicote products comprises of hard wearing release films which offer multiple releases with no fouling transfer from mold surface
- These products include a variety of cleaners, semi-permanent release agents, and sealers.

- Qiicote products contain variable slips which have a high gloss finish.
- Low price of per released part
- The supreme performance of release parts.

Furthermore, various other release agents including an internal mould release agent, water based semi-permanent release, and Silicone shield have been manufactured. Silicon Shield is a modified PVS and can seal up to 100°C. Moreover, it is used to preserve silicon bags.

## DEVELOPMENT OF A NEW CLASS OF COMPOSITE MATERIALS

A Swiss high-tech fermentation company "Evolva" has joined hands with the United States Navy to target their combined efforts on the development of a special, new class of structural composite materials. These composite materials have been developed from a polymer resin matrix which has been contrived from a specific fabrication of Evolva's resveratrol. Presently accessible structural carbon composite materials are usually not suitable to be used for most of the high energy applications which have a potential of catching fire like engine components of vehicles, elevators, fuel tanks, trains, rockets, high buildings, and lithium battery casings, etc.

Resveratrol is a natural ingredient produced by plants when they are subjected to any injury or are under attack of pathogens like bacteria or fungus or facing extreme conditions like high temperatures, dehydrations or any illness. The polymer resin matrix formulated from a distinct formulation of resveratrol is used for the development of this new class of composites. This resveratrol can be manufactured on an industrial scale economically. Resveratrol, with the help of advanced techniques of fermentation and biotechnology, is converted to thermosetting monomers at first; polymerized afterward and then shaped with the help of standard fabrication methods (Guenthner et al., 2014). Preliminary tests of prototype materials developed from Evolva's resveratrol have been carried out to test their performance. Results show that these materials perform much better than prevalent fire resistant materials. Resveratrol thus offers some benefits over fire resistant materials including:

- Polymers made from resveratrol composites are much lighter in weight than aluminum.
- They are halogen free
- They can endure extended intense heat exposures without getting burnt.
- They can tolerate prolonged flame impingement without getting structurally degraded.

Further testing is required to ensure that results remain constant before employing this material for the development of new class of structural composites.

## APPLICATIONS OF COMPOSITES IN MILITARY

Composite materials have application in both land based as well as aerospace industry. Some of the examples are C17, C130, Merlin helicopter and Foxhound armored vehicle. Following are some applications of composite materials specifically in military vehicles.

- Radar panels
- Missile containers
- Avionic racking
- Refueling probe fairings
- Floor panels
- Lightweight transportation boxes
- Avionics cabinets
- Composite doors & covers
- Medical stretcher assemblies
- Radome of radars
- Rotor blade boxes
- Sensor probes
- Add-on armor for military vehicles
- Spall liners
- Armor for light vehicles
- Lightweight protection for gunners
- Protection against landmines

- Protection of weapon stations
- Protection of optoelectronics.

## Composites as an alternative to Titanium

Two organizations namely PROOF Research Advanced Composite Division and the Air Force Research Laboratory are jointly developing materials that are derived computationally as well as manufacturing solutions such as high-temperature polymer matrix composites (abbreviated as PMCs) as an alternative to titanium.

The major benefit of using polymer matrix composites in place of titanium in the aircraft industry is that by using these composites, 40% reduction in weight is achievable. This delivers an additional advantage of a large amount of reduction in annual fuel consumption of air crafts resulting in saving of hundreds of dollars. Using polymer matrix composites also increases the service life of air crafts and improve their fatigue resistance. An additional amount of $750,000 is being paid to PROOF ACD by the Air Force Small Business Innovation Research Program. This funding will be used to transition these technologies with the intention of supporting the Air Force's Technology Program for Integrated Computational Methods for Composite Materials

## New Ballistic Wallpapers

The United States Army has divulged the designing of a new defensive product known as, "ballistic wallpaper" (Fig. 69) which will protect the American soldiers in harsh environments such as war zones (Pridham & Bowery, 2006). Often, at the times of war, army troop uses deserted buildings to take cover for defense as a substitute of constructing their own or burrowed foxholes. The problem arises when these buildings are being fired upon by tank or RPG during the war; when bursts hit the wall, it gets knocked down by exploding, sending shards of glass and hard rocks at the troops sheltering inside.

**Figure 69:** Bullet-proof ballistic wallpapers developed from composite materials

Engineers working at the U.S. Army's Research and Development Centre have been trying different ways to get rid of this problem and now have presented the idea of using "ballistic wallpaper" to fortify such shelters. Walls of shelters can be quickly covered with rolls of adhesive, lightweight ballistic wallpapers. These ballistic wallpapers are formed from Kevlar fibers which are embedded in a flexible polymer film. Although the research center team has carried out blast testing in three different regions (Fort Polk, Eglin Air Force Base, and Louisiana), this wallpaper is still in its research and development phase and does not own an official name yet. It is however expected that soon ballistic wallpapers will be in the market and thus able to save lives.

## Composites for Development of Weapons

It is strongly believed by Military that composite materials are best to make lighter, stronger and long-lasting weapons. The conventional method of making weapons completely from steel is still in practice. However, army has carried out vast research in the field of composites and the associated processes (for example carbon- carbon composites and ceramic matrix) only to find out that not only these materials are lighter than the steel but they can also endure high temperatures to a great deal, which otherwise can be a reason of their system failure. The biggest challenge in constructing weapons from composites is to stick them to the parts of the gun which must be made of steel, barrel for instance. During their initial testing phase, composite materials were enfolded around the steel tube of a howitzer, in a jacket like manner.

Composites allow modifications in their design. Various composite mixtures can be used for various parts of a gun. The flexibility of composites allows us to use more robust and costly composites in precarious areas, while in less sensitive areas, more flexible and less costly composites can be employed. An example can be taken of an effect well known as "coupling" in which certain layers of composite materials are induced. This method, also known as extension twist coupling, has been efficaciously experimented on helicopters to alter the attack angle of its rotors during flight. These composites help in slight twisting of various portions of a gun during shuddering in a manner similar to a rubber band so that the parts may not break.

One of the most advanced illustration of composites development is XM360 1 20mm cannon. To guarantee the proper fitting of the composite jacket over XM360's barrel, its steel core was first compressed by chilling it with frozen carbon dioxide. After that, it was wrapped with the help of thermal plastic in a tight manner. Lastly, the frozen barrel of gun expanded into the composite jacket after getting warmed up. A total of fourteen barrels were manufactured using this method, and each barrel was tested afterward. Few barrels were able to fire as much as 250 rounds. The success of these tests has now made this technology available in the market and ready for use.

Other weapons in their testing phase are the 81 mm mortar tube and a baseplate which have been the subject of research for composites integration. To overcome the problem of excessive heating in mortar tubes, testing of ceramic and metal composites is under progress. Composites provide an additional advantage of the reduced weight of these tubes which makes the job easy for the teams using them. The possible weight reduction by using composites in weapons can be from 90 pounds down to 50 pounds. The weight of only base plate can be reduced down to 15 pounds from 25 pounds. Aluminum matrix composites can also be used for barrel manufacturing (Holloway, 2002). Prototype of XM360 cannon is shown in Fig. 70.

**Figure 70:** XM360 cannon prototype manufactured form Kevlar fiber composites

Efforts are being made to reduce the weight of M1 09A6 / M284 Paladin Cannon bore evacuator by incorporating composite materials in its manufacturing. It is expected that composites will be able to reduce the weight of M284 Bore Evacuator from 200 to 78 pounds. It is also expected that the similar composites will be useful for 155 mm howitzer and should be ready within the next five years.

## The U.S Navy's Carbon Fiber Ship

The largest ship in the U.S. Navy which has been designed using advanced composite materials is the Stiletto. This ship was designed to deliver Special Forces to their respective insertion points speedily. Heavier ships get through strong advancing waves due to their large enough sizes. When power boats are used to drop off the soldiers to their insertion points, the ride usually gets tough as the light weight of power boats causes the boat to bounce through the waves in a rough manner. However, the technology used in M80 has enabled this boat to pass through bumpy sea waves with ease. The efficiency of the vessel is increased by creating an air cushion with the help of bow wave energy. The Stiletto's draught is reduced to mere 3 feet by using this technology even when the boat is fully loaded with tons of equipment and crew members. It further enhances the ride and fuel efficacy introducing very slight drag. The M80 has proved to be a comfortable and speedy ride. It has been built with four Caterpillar engines that are highly adept of a top speed in addition to 60 knots with 500 nautical miles range when loaded fully. Its body has been manufactured almost entirely from carbon fiber which has abridged its weight to mere 45 tons, yet keeping the structural strength of the boat. The Stiletto was used by the U.S. Navy in Colombia in a fight against drugs. This high-speed boat was used in a seizure of around 1800 lbs. of cocaine in the same war. Carbon fiber ship manufactured by for the US army is displayed in Fig.71.

**Figure 71:** Largest US ship constructed from carbon Fiber composites

## The RELAY Station

Fiber Forge has successfully installed its first RELAY station (Fig.72) in Germany, Europe. It is by far the fastest automated tape laying system for composite materials in the world. The RELAY station provides an optimum solution for high volume fabrication of locally reinforced thermoplastics. Fiber forge attempts to hold the advantages of advanced composites made of thermoplastics by using their manufacturing process technology. Using patented customized manufacturing processes have helped Fiber Forge to output high performance and high standard advanced composite components for lesser prices.

Fiber forge starts their processes from the design of a part. Since key parameters like the stiffness and the strength of any part are associated with the fibers, thus allowing composite materials to have many degrees of freedom. As a result, most of the times carbon fiber is being used in abundant quantities as it is one of the stiffest and strongest out of many and helps in manufacturing very light weight parts as per our requirement. These fibers are amalgamated with a matrix material. The choice of the fiber depends mainly on the application for which it is developed, cost issues, the environmental necessities and the performance requirements.

**Figure 72:** Army RELAY Station constructed from composite components

The process of manufacturing is initiated with a unidirectional tape of prepreg. According to Fox-Rubin of Fiber Forge, for the majority of parts, any fiber and any thermoplastic resin can be transformed into a tape material. At first, THIS tape is placed in a flat pre-form in a pre-arranged placement and thickness of fiber by using automatic RELAY system of Fiber Forge. This pre-form of tape is known as a Tailored Blank, a term widely used in auto industry, for the process of tailoring the properties of steel sheets prior inscribing them in the forging process. After the development of the tailored blank, all of its layers are joined by passing it through a two-stage consolidation process. In the first stage, the blank is heated under pressure, and in the second stage of consolidation process, the blank is cooled down, though still under pressure. In this manner, a consolidated panel is created.

After the consolidation process, the blank is transported to a thermos forming process. As mentioned by Fox-Rubin, this forming procedure is quite rapid in nature. The molten form of resin is picked up, and then it is frozen into its final part shape. Beforehand fibers were designed and placed in a proper manner according to the requirement of the desired shape, thus, making the fibers go in the direction where they are supposed to be.

## REPAIR OF COMPOSITE STRUCTURES

The development and maintenance of advanced composites and structures (such as carbon fiber, fiberglass, and Kevlar composites) is a great challenge for Army. Two main organizations associated with composites development and maintenance are the U.S. Army Aviation and Missile Research, Development and Engineering Centre (AMRDEC) which has the facility of Prototype Integration generally abbreviated as PIF. Initially, the concept of using advanced composites in flight structures of air crafts was alien to the army as soldiers lacked the necessary skills to develop and maintain/repair composite components. Afterward, Prototype Integration Facility was developed which can successfully design and validate the repairing process. The facility further trains the soldiers so that they may undertake those repairs. In this repair course, soldiers are given a synopsis of the materials, processes, and tools which are used in repairing of advanced composites. A good thing about the course is that it is applicable to all aviation platforms. The PIF, in the previous two and half years, has trained over 250 soldiers as well as civilians in advanced composite repair technologies and other fundamentals of composites.

In February 2014, the PIF was granted the Army Aviation Association of America Material Readiness Award for their contribution to the extensive implementation of composites. Till this date, the PIF has maintained a direct and intense effect on aviation and is the center of aviation activities due to its vast experience in advanced composites.

## PRESENT AND FUTURE CHALLENGES AND SOLUTIONS

While working with composites, it is important to understand the effect which each part causes on the other. While dealing with structural composites, the mere fact that the design looks good cannot be taken as a guarantee that the part will perform well. For instance, the reinforcements could be wrong. Equally, if any part is well designed structurally and bears all of the loads and tensions well during analysis, but lacks a decent shape and design, presents a problem as well.

As mentioned previously, the biggest challenge while using composites in the manufacturing of weapons (for example, a gun) is to make them stick to the steel portions of the gun. The stiffening property of the gun helps in reducing vibrations; therefore to ensure high targeting accuracy as well as to reduce vibrations, it is necessary for a gun to demonstrate such "stiffening" properties (Harry, 1958).

Another feature which must be taken care of is the dynamic strain. The dynamic strain is the probability of the breaching of the material. Failure of dynamic strain can prove to be dangerous, for example, during one testing, failure of dynamic strain caused the muzzle end of the gun to break apart. As a result, the end of gun flew downrange.

Another beneficial property of composites is the flexibility they allow in their designs. However, it is vital for companies manufacturing composite materials to keep their customers in the loop from product design to its development cycle.

The experience of developing composites from different combinations of materials as well as their analysis needs much attention. Reliable CAD / CAM analysis is still lacking. For instance, the composite analysis tool for crash energy absorption is not up to the standard of its metal counterpart analysis tool. The material properties of composites are not same in every direction which makes them quite difficult to model. Thus, it is understandable that software community needs more time to fully develop advanced analysis tools to be used by designers.
Another issue faced by the composite industry is the high temperature. Although composites are made to tolerate high temperatures, still most of them are unable to tolerate temperature as high as engine combustion, for instance. More research should be carried out to find composites which can work in extreme temperatures. Composites have been used in the manufacturing of engine blocks for vehicles used for automotive racing, but even so, they have a lifetime of only one race in those vehicles, which leaves the room for expansion in the use of reliable composites in elevated temperature locations (Allison & Cole, 1993).

Lately, a lot of progress has been made in the field of resin development from advanced composite materials. The toughness and relatively lower cost of resins are the reason they are expected to be used more than thermosets in wheeled vehicles. Some companies are now developing a variety of thermoplastic resins. These thermoplastic resins are designed, developed and afterward used in large quantities (approximately millions of pounds per year) for various purposes, injection molding for instance.

For the development of structural parts, these same kinds of resins can be joined with continuous fibers. A major focus is not on developing new composite materials, but rather on devising processes for converting already developed composite materials into usable parts.

## WEAPONS UPDATES

It is mostly believed by engineers of the U.S. Army Armament Research, Development, and Engineering Center that composites can aid the Army to have lighter, stronger and sturdier weapons. Like the old times, most of the weapons are still manufactured nearly completely from steel, as reported by an article published by the Army News Service. Application of heat resistance is of utmost important for the army to develop their weapons because often extreme heat causes weapons system to fail. Some composites are more heat tolerant and therefore can withstand high temperatures; moreover, an added benefit is their light weight properties.

Army has used composites for the development of some part of XM360 120mm canon. As mentioned earlier, composite was used in the manufacturing of jacket of canon which was fitted on the barrel of the XM360. Fourteen barrels were manufactured in this way and tested afterward. Successful tests have now led this technology ready for use in the market. Other weapons which are now the subject of research for composites integration are the 81mm motor tube and base plate. Testing is soon expected to be carried out as reported by the Army. According to Littlefield, a soldier serving in the Army, composite materials made of ceramics and metals are under test to help to deal with the problem of excessive heating in motor tubes. Composites will also help to lessen the weight of these mortar tubes. The weight of base plate can be reduced to 15 pounds from 25 pounds, thereby, making them easy to carry.

Previously, composite technology has been used in the development of M256 cannon of the Abrams tank (Littlefield et al., 2008). Now, using the same technology, work is in progress to reduce the weight of the M109A6/M284 Paladin Cannon bore evacuator. At present, the weight of M284 Bore Evacuator is around 200 pounds. It is expected that composites will reduce this weight to approximately 78 pounds. These newly developed composites will be ready for consumption in almost five years, and may also be available for use in 155mm howitzer.

## REFERENCES

1. Hogg, P. J. (2006). Composites in armor. Science, 314(5802), 1100-1101.
2. Cohen, M., & Israeli, A. (1998). U.S. Patent No. 5,763,813. Washington, DC: U.S. Patent and Trademark Office.
3. Szczygieł, B., & Kołodziej, M. (2005). Composite Ni/Al 2 O 3 coatings and their corrosion resistance. Electrochimica Acta, 50(20), 4188-4195.
4. Mangino, E., Carruthers, J., & Pitarresi, G. (2007). The future use of structural composite materials in the automotive industry. International Journal of vehicle design, 44(3-4), 211-232.
5. Molitor, P., Barron, V., & Young, T. (2001). Surface treatment of titanium for adhesive bonding to polymer composites: a review. International Journal of Adhesion and Adhesives, 21(2), 129-136.
6. Jones, R. M. (1998). Mechanics of composite materials. CRC press.
7. Yang, W., Song, L., Hu, Y., Lu, H., & Yuen, R. K. (2011). Enhancement of fire retardancy performance of glass-fibre reinforced poly (ethylene terephthalate) composites with the incorporation of aluminum hypophosphite and melamine cyanurate. Composites Part B: Engineering, 42(5), 1057-1065.
8. Tucker, R. G. (2012). U.S. Patent Application No. 13/490,631.
9. Kritchevsky, G. R., Gregor, J. A., Gruendig, M. W., Sellers, G. J., & Liss, B. (1987). U.S. Patent No. 4,678,699. Washington, DC: U.S. Patent and Trademark Office.
10. Guenthner, A. J., Davis, M. C., Vij, V., Yandek, G. R., Lamison, K. R., Reams, J. T., ... & Mabry, J. M. (2014). New Approaches to Maximizing Thermo-oxidation Resistance of Polycyanurate Networks (No. AFRL-RQ-ED-VG-2014-230). AIR FORCE RESEARCH LAB EDWARDS AFB CA.
11. Pridham, B. J., & Bowery, M. J. (2006). U.S. Patent No. 7,050,286. Washington, DC: U.S. Patent and Trademark Office.
12. Holloway, S. R. (2002). U.S. Patent No. 6,482,248. Washington, DC: U.S. Patent and Trademark Office.
13. Harry, D. (1958). U.S. Patent No. 2,845,741. Washington, DC: U.S.

Patent and Trademark Office.

14. Allison, J. E., & Cole, G. S. (1993). Metal-matrix composites in the automotive industry: opportunities and challenges. JOM Journal of the Minerals, Metals and Materials Society, 45(1), 19-24.
15. Littlefield, A., Hyland, E., & Keating, J. (2008). Prestressed carbon fiber composite overwrapped gun tube (No. ARAEW-TR-08018). ARMY ARMAMENT RESEARCH DEVELOPMENT AND ENGINEERING CENTER WATERVLIET NY BENET LABS.

# CHAPTER 8

# APPLICATIONS OF COMPOSITE MATERIALS IN MINING INDUSTRY

## INTRODUCTION

The concept of composite materials is based on the idea of harnessing the customized properties of both fibers and reinforcement to develop high-performance materials. Fibers are typically responsible for providing strength and stiffness while matrix provides toughness and ductility to a composite material. Distinct combinations of various fiber-reinforcement species can be employed to tailor the properties of composites such as stiffness, mechanical strength, fatigue life, weight, acoustical insulation, heat resistance, and resistance to corrosion and wear.

Presently, tailored composite materials are being extensively used in automotive, aerospace, petrochemical, and metallurgical industries. Composite materials are also being employed in electrical, biological, civil, marine, and structural mining applications (Markgraaf, 1995).

According to the latest reports, the European mining industries are taking great interest in mining and metallurgical applications of composite materials (Waterman, 1991). However, without the use of nature's best composite material, wood, which is a natural support material in structural applications, the modern mining industry would have been struggling to make its way towards the sustainable development of reinforced concrete. Recently, the revolutions in the composite sector have resulted in the utilization of particulate reinforced for mining applications. Moreover, modern composites developed from the glass fibers and different fire-retarding polymers are being abundantly used in the USA for mining applications such as rock bolting, pipework, ventilation ducting, underground transport, equipment housing and conveyor belting fans.

Many researchers have discussed the novelties in composite materials and their usage in commercial sectors. This chapter, however, focuses on the applications of composite materials in mining and metallurgical industry. The aim of this chapter is to shed light on the excellent characteristics of composite materials. The modern composites have replaced traditional ceramics, metals, and plastics because of their flexible design, and ability to provide customized material characteristics. The developments in the composite manufacturing technology have revolutionized the mining industry. Composite based light weight mining equipment have reduced transportation costs which result in efficient and cost effective mining processes. The composite material is a mixture of two or more distinct material phases, with discrete interfaces separating the constituent elements, combined in a three-dimensional way to provide synergistic physical and mechanical properties which are superior to those of the individual components (Carter & Paul, 1991).

**Figure 73:** Composite based feeder drive system for mining operations

The composite materials are an essential part of the modern functional materials. As compare to the traditional structural materials, composite materials provide customized set of properties for load bearing applications. Unique characteristics of composite materials include robust design, high specific stiffness and strength, excellent fatigue resistance and, customized integration of structural components. Traditional materials cannot fulfill the criteria for satisfying a range of structural, thermal and chemical applications. Composite materials, however, serve as multidimensional material systems for manifold industrial applications. Modern composites are among the essential raw materials to manufacture high-performance products for mining applications (Jones, 1998). In addition to outstanding mechanical and structural characteristics, composite materials also possess remarkable insulating properties for thermal, electrical and acoustic applications. Considering the immense potential of composite materials for mining applications, they are widely used in coal mines.

**Figure 74:** Welded Composite wire-mesh rock bolting for coal mine

## BACKGROUND

Compositematerialsaretypicallypublicizedas"Novel"or"Revolutionary" materials. However, composite materials have been utilized by the human race for centuries. Wood is one of the first natural composites which have been utilized for structural applications by ancient people. Moreover, the famous fiber reinforced plastics (FRPs), are approximately 50 years old and have been employed in various applications since their origin. Composite materials are an excellent choice for manufacturing corrosion resistant and chemical resistant industrial structures (De-Beer, 1966). Besides their excellent chemical properties, composite materials also possess other unique properties such as high strength and stiffness, low weight, low cost, high thermal resistance, customized properties, ability to be formed into complex shapes, flexible manufacturing processes, tailor-made surface characteristics and beautiful aesthetic features. Considering the unique structural and thermal properties of composite materials, they are an excellent choice for mining applications.

Carbon fiber reinforced phenolic resins and modified thermoplastic resin (Modar from ICI) are ideal materials for mining applications because of their excellent load bearing and fire retarding characteristics. It has been reported that phenolic-carbon fiber reinforced composites are a suitable choice for mining structures including elevated walkways, racks, floor support structures, handrails, and ladders. Composites are conveniently pultruded to form standard profiles of the structural components (Ashby & Jones, 1985). Reports have also shown the use of pultruded composite structures in European coal mines for flooring, geotextiles, heap leaching pads and liners for ponds. Composite materials can also be used for rock-bolting in underground mines because of their flexibility to be installed in variable sizes. Literature has shown the usage of filament wound carbon fiber reinforced plastics for installation and repair of underground pipelines and ducts (Harris, 1991). Currently, laminated aluminum is being used as a light weight housing material for hydraulic equipment. However, comprehensive research is also being carried out to evaluate the performance of composite materials for the housing of hydraulic equipment such as roof-support props. In British Columbia, floating pump platforms made of pultruded Polymer matrix composites (PMCs), are being utilized for copper leaching process by Gibraltar Mines Ltd, which is prominent mining company in Canada. Copper leaching process involves the spraying of sulfuric acid onto the copper ore, which results in the collection of copper leachate into a composite lined holding pond.

As compared to steel structures, composite based underground walkways and platforms are lighter, flexible and easy to install. Moreover, these structures are designed to float on the water surface without sinking into the water at different water levels. Other industrial and mining applications of composites include abrasion-resistant material components, pressure vessels, FRPs based hydro cyclones and pressure vessel safety attachments (Crowson, 2016). Carbon fiber reinforced ultra-high-molecular-weight (UHMW) polyethylene composites are used to manufacture ultra-strong, tough, lightweight, shock-resistant and mechanically safe hoisting cables and safety helmets. Currently, composites are being used in industrial applications such as wear-resistant gears, cutting-tool inserts, heat exchanger tubes and kiln lining materials. Further mining uses of composite materials include sludge pump housings, hard and wear resistant tips of rock-drills, corrosion, abrasion and erosion resistant pump shafts, and components for applications in corrosive environments. Some mining components are produced by a commercial manufacturing process called Direct metal-oxidation processing (DIMOX).

The DIMOX process involves the production of reinforced alumina by using particulate (SiC) reinforced preforms. The modified preform is retained in a furnace which contains molten aluminum metal. The molten metal diffuses into the preform, and oxidation of the aluminum alloy takes place under controlled conditions; resulting in aluminum oxide accumulation around SiC grains. The DIMOX process is typically used for manufacturing of a wide range of wear resistant composite parts for use in mining and power generation. Typical composite products include cyclone apexes and liners, scraper blades, valve parts, pump liners, nozzles, elbow wear backs, deflector blades and chute liners for fuel transportation and coal-preparation systems (Tenenbaum, 2009).

Apart from FRPs, there are some other categories of composite materials which are used to manufacture components for the mining industry. Carbon/Carbon (C/C) composites (carbon matrix material reinforced with carbon fibers) are among the most popular high strength and stiffness materials for mining applications. Another important category of high-performance composites include composites manufactured from silicon nitride ($Si_3N_4$) matrix and silicon carbide (SiC) fibers. C/C composites are capable of withstanding extreme temperatures, and high impact loads due to energy absorption capability of their exceptional 3D structure. These composite materials usually exhibit high fracture toughness, low thermal shock sensitivity, and pseudo-plasticity comparable to that of FRPs. Moreover, C/C composites retain their mechanical and thermal characteristics up to 2000°C (Park, 1992). Although conventional C/C composites show negligible resistance to oxidation at high temperatures, novel fabrication techniques have been developed to make these composites oxidation resistant at elevated temperatures. Presently, continuous-fiber $SiC/Si_3N_4$ Composites are being utilized in various mining applications such as waste incineration systems, heat recovery equipment, and manufacturing of wear-resistant structures.

## COMPOSITE APPLICATIONS IN MINING INDUSTRY OF USA

Applications of composite materials in the mining industry of USA are increasing steadily with time. Some of the mining uses of composite materials include chemical resistant pipes for refining of platinum, cladding for insulation of chilled water containers, and ducts for ventilation systems.

The use of composite materials in the USA is well documented. The most common uses of composite materials are discussed below:

- The coal mining industries of USA utilize glass fiber reinforced polymers (GRPs) to manufacture composite pipes for the transport of slurries, acid water, and fresh water. GRPs have replaced steel products because of their excellent corrosion resistance characteristics during the handling of acids. Successful installations of composite products can be witnessed at leading American industries such as Carbon Fuel, Eastern Associated Coal, North American Coal and Consolidated Coal.
- Currently, bottom ash in the coal mines is being transported via GRP-composite pipes by most of the companies. The ash typically consists of 12% iron oxide ($Fe_2O_3$), 40% silica ($SiO_2$),

16% alumina ($Al_2O_3$), 22% calcium (Ca), 2% magnesium (Mg), 1% sodium (Na) and traces of other materials. Besides the provision of cheaper piping products, the lifecycle of composite materials is significantly longer than the steel installations. Philip Sporn Plant and Central Operating Co. of West Virginia use composite pipes for the transport of natural ash at 10% concentration. The lifetime of the pipeline is estimated to be indefinite at the concentration equal or less than 10%.

- The Hodge-Union Texas project, installed in Oklahoma State, is considered to be the largest GRP-composite installation around the globe which contains about 150,000 m long pipe. The composite pipe was fabricated to combat the corrosive environment and paraffin attacks.
- A GRP-composite pipe (800m long) has been installed at a paper mill in Wisconsin for transport of waste water. Previously, stainless steel pipes were used for the movement of the waste water. However, steel pipes faced severe corrosion due to the presence of sodium hydroxide in the system. Presently, GRP pumps and pipes are being broadly used for corrosion resistance applications in the USA.

## COMPOSITE CHARACTERISTICS FOR METALLURGICAL AND MINING APPLICATIONS

### Excellent Mechanical Properties

Composite materials have a high strength to weight ratio due to their lighter weight and outstanding mechanical properties. The density of GRPs is typically 20-25% the density of plain carbon steel. GRPs are also around three times lighter than aluminum alloys. Moreover, the value of mechanical strength of the composites may exceed that of the plain carbon steels. Carbon fiber composites and organic fiber composites possess higher strength and lower density than the GRP-composites. High specific strength and specific modulus values of advanced composites are primary parameters for their applications in load bearing structures.

## Chemical and Corrosion Resistance

Composite materials exhibit remarkable resistance to corrosion and chemical attacks. There is a broad range of corrosion resistant and acid resistant composite materials, such as composites manufactured from glass fibers reinforced in phenolic resins. Such composites can be used to construct pipelines, valves, pumps, stirrers and container for prolonged storage of chemicals (salts, acids, esters, etc.).

## Excellent Fatigue Resistance

Composites display excellent fatigue resistance properties under cyclic loading conditions. Fatigue fracture occurs due to the nucleation and propagation of cracks (voids) during variable loading cycles. Fatigue failures may occur catastrophically in metallic materials without any prior indication of the crack. However, the composite materials deal with the cracks differently due to the presence of non-directionality in the composite structure. Inhomogeneity in the composite structure caused by different fiber orientations serves as dislocation pinning site, which prevents the material from catastrophic failures. Fatigue limit of the metals is around 40% ~ 50% of their ultimate tensile strength while fatigue limit of carbon fiber composites is approximately 70-80% of their original strength. Additionally, FRPs also possess high fatigue resistance to acoustic vibrations.

## Self-Lubrication and Wear Resistance

Most of the polymeric materials have excellent self-lubricating characteristics due to the presence of weak hydrogen bonding or Van der Waals forces between their molecular chains. Additionally, polymeric resins have low coefficients of static and dynamic friction which make them excellent self-lubricating compounds. Filler compounds are added to the polymeric resins to tailor their mechanical and self-lubricating characteristics. Fiber reinforced polymeric composites are suitable materials for use in mining applications where low wear and friction is desired.

## Thermal and Electrical Insulation of Components

Glass fiber reinforced polymers (GRPs) are excellent materials for electrical insulation of mining-motor parts. Composites are also immune to electromagnetic radiations and radio waves. GRPs also exhibit resistance against ablation and underground, irradiation.

### Extraordinary Thermal Resistance

Heat resistance of carbon fiber reinforced metal matrix composites (MMCs) is remarkable. MMCs have shown superior thermal resistance characteristics than most of the metals (Hanzawa & Nakano, 2002). It is a fact that, steel becomes unstable over 700 °C while most of the other metals also show phase transformations at elevated temperatures. On the contrary, ceramic matrix composites are considered to be the appropriate materials for thermal resistance because of their higher decomposition temperature. Alumina fibers and silicon carbide fibers in conjunction with ceramic matrix can form composites which can endure temperatures up to ~1400°C without any degradation. Ceramic matrix composites are lighter in weight as compared to metallic alloys. They can be used for high-temperature smelting application in the mining industry.

## POTENTIAL MINING APPLICATIONS OF COMPOSITES IN SOUTH AFRICA

South Africa (SA) is one of the major countries rich in minerals. Mining industry of SA has been utilizing conventional materials for decades. Recently, there has been an industrial shift towards composite materials in mining industries. There are many applications of composite materials in the mining industry of SA, some of which are listed below:

- Different kinds of light weight mine props
- Insulated chilled water pipes
- Lightweight hoisting ropes and skips
- High capacity and light weight gas cylinders for breathing and firefighting
- Anti-corrosion air handling equipment
- Slurry transport
- Ventilation pipes
- Thermal insulation claddings

- Hydropower pipes
- Corrosive fluid storage and handling
- Corrosion resistant gratings and cable trays
- Blast protection structures
- Water hydraulics
- Air actuators

## Mine Props

Mine props are usually fabricated from steel due to its specific characteristics such as low initial costs, fine mechanical durability, and high blast resistance. However, there are some negative aspects of using steel as props material which includes poor corrosion resistance, high weights, high processing costs and inflexible manufacturing techniques. Composite materials have gained immense recognition as alternative materials for steel due to their high strength to weight ratio and excellent corrosion resistance. Composite mine props for underground mining operations are shown in Fig.75.

Figure 75: Composite mine props for underground applications

Till date, various attempts have been made to manufacture the mine props from substitute materials. However, the use of composite materials for manufacturing of props has not been explored in the past. Presently, research is underway to shift the props manufacturing from steel to composites entirely.

## Chilled Water Transport

Steel pipes are extensively used for transport of chilled water in underground mines. Steel pipes are great carriers of chilled water, but their insulation is required to prevent the cooling loss. Moreover, steel pipes are not fire resistant, and their cost is also very high. Thermoplastic (PVC, PE) pipes are suitable candidates for providing insulated chilled water piping. However, thermoplastics are poor fire retardants, which makes them unsuitable for the underground piping systems. Fiber reinforced thermoset phenolic pipes can be used as an alternative for steel pipes due to their enhanced fire resistance and thermal insulation characteristics. Moreover, composite pipes are also mechanically stiffer and stronger than plastic and some steel pipes. It is essential to evaluate the compatibility of the composite piping systems with the existing mining installations to avoid any extra costs. The benefits of the new composite systems are:

- Extra insulation is not required (i.e. cladding cost is omitted)
- Inherent corrosion resistance of the pipes saves a lot of money (used for replacement of steel pipes)
- The light weight of the pipes makes their transportation and installation easy.
- The composite pipes would have fundamentally low pumping losses because of their excellent internal surface finish.

## High-Pressure Piping Systems

The strength and toughness of the composite materials are comparable to steel and other alloys. A composite piping system is suitable for the transportation of high-pressure fluid (water) for generation of Hydro Power. The inherent corrosion resistance of the composite material can prove to be helpful if the source water for energy production is untreated. The corrosion resistant composite pipes can outperform steel pipes in most of the cases. On the contrary, the development of composite systems requires the on-site production of the composite pipes. Additionally, composite manufacturing of conforming components for fittings and joints will also be needed for compatibility with the composite piping systems.

## Corrosive and Chemical Fluid Handling

Most of the metals show vulnerability to chemical and corrosion attacks to some extent. Therefore, it is imperative to develop a class of materials which can perform well without any degradation. Composite materials have shown tremendous potential for chemical and corrosion resistance applications. The excellent resistance of the composites materials to almost any chemical or corrosion attack makes them the ultimate corrosion resistant materials (with no competition). Other alternate materials have failed miserably to serve the purpose. To completely replace the conventional materials, FRP industry still needs to perform well economically.

The successful application of the composite materials in the field of corrosion resistant piping system depends on the collaboration between the clients and resin manufacturers. The careful selection of a resin system is essential for the production of high-performance composite materials.

## Water Hydraulics

Nowadays, it is quite usual to use the GRP composites for water hydraulic systems. Composite hydraulic cylinders have many advantages over conventional cylinders such as light weight, and high corrosion resistance from highly corrosive untreated water. The lightweight composite cylinders provide the efficient and profitable water transport solutions. Apart from light weight and corrosion resistance, composite cylinders are also wear-resistant and abrasion resistant, if abrasive particles are present in the water. Moreover, the abrasion resistance can be improved by the addition of wear resistant lining of ceramic-based composites.

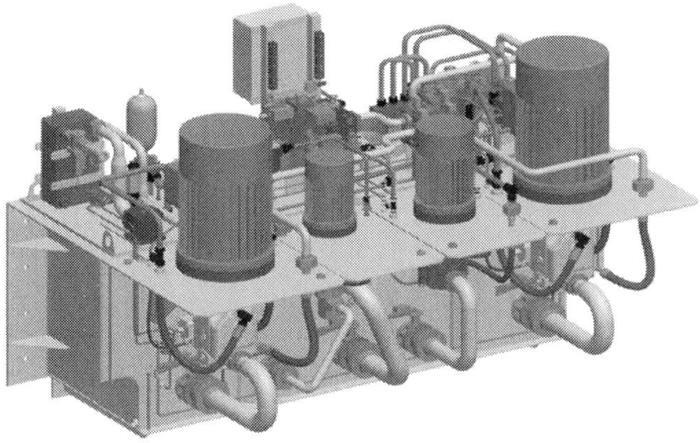

**Figure 76:** Hybrid composites based water hydraulic system

The biggest problem of thermoplastic GRP composites is their small fire resistance capacity. This issue can be rectified by using phenolic resin matrices in conjunction with glass fibers. The best combination of fiber-resin content can be mixed to develop composite materials for hydraulic systems.

## Slurry Transport

Use of composite materials for the transportation of mining slurries has been expanding in the US for the last few years. The unique composite technology is the solution for manufacturing of cost effective transport pipes and containers. However, this technology for production of slurry transport mechanisms will mature gradually with time.

## Skips

The deadweight (mass) of the empty skips directly influence the performance of principle transportation systems in mines. Reduction in the deadweight of the skips is essential for reducing the energy consumption, which ultimately enhances the productivity of the process due to increased hoisting speed of lightweight skips. Composite skips are not being widely manufactured due to the lack of proper guidelines for mining industries. On the contrary, composite skips are quite easy to produce as compared to other mining equipment. South Africa has developed the manufacturing systems for composite skips, and extensive research is being carried out in this field. To implement the commercial production methods of skip construction, certain technical requirements need to be addressed, which include enhanced life cycle, and cost effectiveness of the composite skips as compared to conventional metallic systems. The benefits of the composite skips are not limited to the amplified productivity only. Other benefits of the composite skips include:

- High safety factors and enhanced payloads
- Lower maintenance costs due to high corrosion resistance
- Low wear on cables and hoists due to mass saving
- Improved working depths
- High strength to weight ratio with almost 40% mass reduction (i.e. same strength at 40% lower mass as compared to conventional steel skips)

## Lightweight Breathing Equipment

Composite materials are famous for the production of lightweight Breathing Apparatus (BA) cylinders in the US. The main advantages of composite BA cylinders include:

- Greater storage capacity at lower cylinder weights
- Resistance to catastrophic failure
- Anti-corrosion characteristics
- High fire resistance

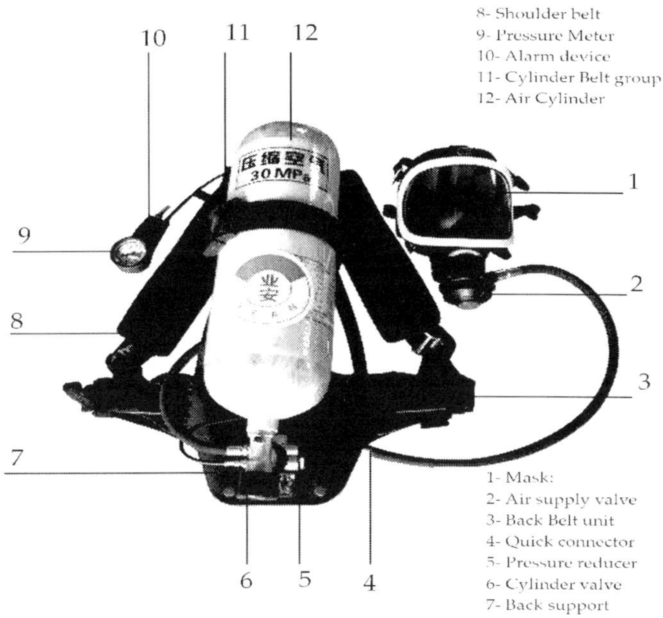


**Figure 77:** Lightweight oxygen cylinders for underground mining operations

Application of composite materials for manufacturing of breathing cylinders would require in-depth investigation and development. The necessary technology is accessible, and there seems no difficulty in the elaboration of this technology.

## APPLICATIONS OF COMPOSITES IN COAL MINES

There are various uses of composites in coal mines. FRPs are most commonly used materials for the production of coal mine equipment (Wang et al., 2013). Major applications of composites in coal mines are illustrated in the following text.

### Underground FRPs Pipelines

Acidic erosion and salt water corrosion are common in the metallic piping of underground coal mines which results in premature failure of the metal pipes. Pipes manufactured from FRP composites contain a resin which can form corrosion resistant film which acts as a shield against the corrosive media. Moreover, the composite pipes have a smooth internal surface which provides strong resistance to corrosion and least resistance to flowing fluids. In the mid-1970s, the FRP-composite pipes were first utilized in coal mines for the transport of coal, air, and water in the USA, Japan, and West Germany. A significant number of experiments have verified that the development and utilization of glass fiber reinforced plastic (GRP) pipelines can accelerate the production rates of coal mines by providing durable, cost efficient and sustainable transporting systems. In the 1990s, China produced high strength and fire resistant pipeline for coal mine applications. The FRP-composite pipes possessed superior mechanical (strength & stiffness) and thermal (Fire & temperature resistance) properties than their previous versions. Expansive research and experimentation resulted in the development of advanced processes for manufacturing of FRP-composites for piping applications in coal mines. Currently, composites are being utilized to manufacture pipes for coal mines on a large scale. These composite pipes provide a reliable and sustainable solution for coal mine transportation systems (Yungui et al., 2001; Pickering, 2006).

## Coal Mine Shafts

The coal mine shafts operate in extreme environmental conditions which include high humidity, high susceptibility to acid-base erosion and high wind speeds. These harsh environmental factors affect the performance of the coal mine shaft by damaging its structure. Coal mine shafts have been constructed from steels and other metals for decades. Oxide films are formed on the metallic surface in the presence of air and humidity. The localized damage (erosion) of the oxide film results in the accelerated corrosion of the steel pipes which may fail catastrophically. Since the mid-1970s, various studies have been conducted to develop the anti-corrosion materials for coal mine applications. It has been reported that the usage of composite materials has proved fruitful for addressing the issues related to erosion, corrosion and fire damages. Moreover, composite coatings on metallic structures are also being developed for providing fire retardant and anti-corrosion pipes.

In the 1980s, some research groups commenced their study on the development of anticorrosion coal mine shafts from composite materials. The experimental study consisted of a research phase followed by a testing phase for both composites and metals. Anti-corrosion and wear resistance tests were carried out for composite shafts, and their results were compared with those of steel shafts. Composite shafts successfully outperformed conventional steel shafts regarding corrosion resistance wear resistance and mechanical strength. Lately, the research groups researched and developed some advanced FRP-composites which were utilized as the corrosion resistant structural material for coal mine shaft. In 1994, the development and publication of the utility model patents on the fabrication of flame retardant and antistatic GRP-composites opened up new ways for the development of sustainable anticorrosion mine shaft equipment (Nanni et al., 2014).

**Figure 78:** Composite coal mine shafts with metallic supports

## Glass Fiber Reinforced (GRP) Composite Bolts

Glass fiber reinforced plastic (GRP) Composite bolts are being abundantly utilized in coal mine walkway supports, exhibiting inherent mechanical and physicochemical properties. Unique characteristics (antistatic, flame retardant and anti-spark) of composite bolts are responsible for providing safety and improved efficiency in underground coal mine operations. Composite bolt has low thermal expansion coefficient (comparable to concrete), and they exhibit excellent immunity against UV radiations in underground mines. Ultimately, composite bolts undergo slow aging due to their low relative mass density (1.5~2.0 t/m$^3$) as compared to high mass densities of steel bolts. Bolts manufactured from GRP-composites are being widely used in coal mines of many developed countries such as the USA, New Zealand, Germany, Britain, and Australia. There is a documented evidence in national laws and regulations of Australia regarding the application of composite bolts in roadway supports. According to the US government statistics, the number of American coal mines utilizing glass fiber reinforced plastic bolts has been increasing abundantly as the number of cigarettes (140 million) escalated in 2002. In recent times, China has been spending abundant resources to develop the advanced technology for manufacturing of composite bolts and pipes. The novel FRP bolts have strong adhesive and tensile strength as compared to metallic bolts. Enhanced flexibility of composite bolts is essential for meeting the requirements of coal mine operations (MA et al., 2006).

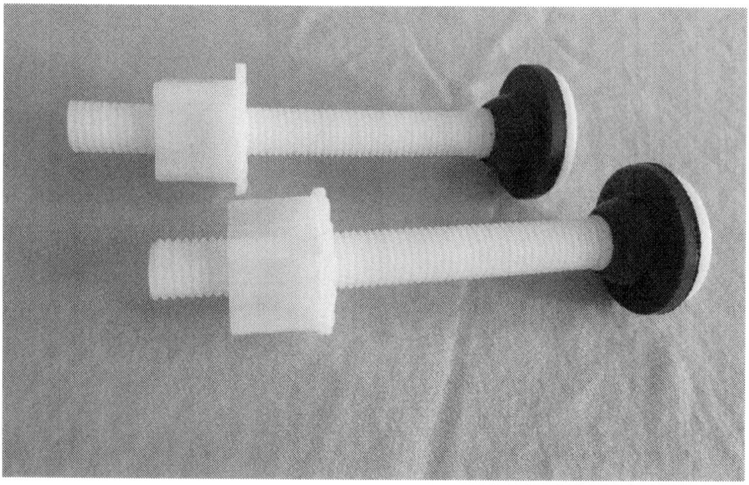

**Figure 79:** Non-corrosive composite bolts for installation in corrosive environments

## Miscellaneous Composite Applications in Coal Mines

There are many structural and non-structural components in the coal mines which include limiting stopper, control button, explosion-proof electrical enclosures, explosion-proof electronic switches, electrical control box, junction box and explosion-proof lamp. Conventionally, these components are manufactured from steel, plastics, and concrete. However, the advent of composite technology has revolutionized the mining industry by providing cost efficient and durable products.

Composite retrofit technology has replaced conventional repair technologies due to its high efficiency and sound processing techniques. Composite retrofit technology is being broadly utilized for repair of water pipelines in underground coal mines. Carbon fiber and glass fiber are usually employed for repair of structures in the coal mines. Carbon composites are considered to be the trend setters in the market due to high values of elastic modulus and tensile strength of the carbon fibers (Laney, 2002). The idea behind the use of composite repair technology is to harness the directional mechanical properties of fibers and strong bonding strength of the resins (Mohitpour, et al., 2002). Composite material retrofit technology is mainly based on the use of carbon fibers and uncured prepregs for basic rehabilitation of the structures. The cured prepregs in conjunction with carbon fibers provide necessary strength and support to the structures. Carbon fiber composites are suitable materials for replacing conventional repair materials for retrofit of pipes and containers in coal mines. Carbon fiber composites possess high inter-laminar shear strength, low water absorption, exceptional corrosion resistance, superb anti-aging performance, and high thermal resistance characteristics, i.e. thermal expansion coefficient of carbon fiber composites is superior to that of steel. The retrofit layer of carbon fiber composite on the outer surface of the pipes share the load experienced by the pipe and reduce the localized stress concentration to strengthen the pipe against extreme loading conditions (Shamsuddoha et al., 2013).

Apart from mine hoisting cages, the liners for car trunk in an underground mine are also developed from carbon fiber composites. Carbon fiber composites are ideal materials for manufacturing highly efficient load bearing components due to their high strength to weight ratio (four times lighter and ten times stronger than steel). Therefore, carbon fiber composites are an excellent source to provide lightweight yet efficient structural components for coal mining operations Buckley & Edie, 1993).

## Applications of Rubber-Based Composite Materials in Coal Mine

- Transportation of mining products is one of the essential processes in the mining industry. Belt conveyor machines are commonly utilized for underground and on-ground transportation of mining goods. The conveyor is typically manufactured from composite materials consisting of cord fabric and rubber matrix.
- Pressure air hoses and water delivery hoses which are utilized in coal mines (both underground and surface sites) are produced from fiber cloth reinforced rubbers.
- The tape hairdryers are commonly used rubber based composite products in coal mines and metallurgy mines for ventilation purposes. Moreover, rubber based mineral tape hairdryers are usually installed in the mine tunnels for vertical or shaft ventilations.
- Lightweight hoisting rope for coal mines is also made of a rubber-based composite material which provides high strength, corrosion resistance, abrasion resistance, flexibility and enhanced lifting efficiency.

## The Applications of Cement-Based Composite Materials in Coal Mines

Reinforced concrete has been under human use for centuries. Concrete and reinforcement are two distinct materials. The reinforcement material is designed to provide high tensile strength while concrete possesses high compressive strength due to its unique formulation chemistry, i.e. formulated from cement and aggregate of gravel and sand. Reinforced concrete is a customized material which can sustain high tensile and compressive loads due to the combined characteristics of both concrete and reinforcement. Currently, the reinforced concretes is being broadly utilized in coal mine industry. In recent times, in-depth studies are being carried out on types of components and structures developed by reinforced concrete. The design techniques and calculation theories are evolving rapidly due to the developments in the composite technology. Fiber concrete is a class of composite materials which consists of concrete as base material mixed with a range of fiber (reinforcement) materials. The primary purpose of integrating fibers into concrete is to improve its tensile strength and decrease its brittleness. Commonly utilized reinforcement fibers include glass fibers, steel fibers, asbestos fibers, synthetic fibers and carbon fibers. With the expansion of the research and development work on composite materials, fiber reinforced concrete has, efficaciously, been utilized in coal mines (Pang & Zhu, 2007).

Fiber jetting concrete mainly known as steel fiber sprayed concrete (SFRC) is another unique variant of reinforced concrete which is commonly used for roadway supports in coal mines. The SFRC or fiber jetting concrete is a particular type of the composite materials which consists of steel fibers reinforced in a sand spray base. In SFRC composites, steel fibers are distributed in two dimensions in a jet plane. Studies have shown that the fraction of steel fibers which are parallel to the jet planes are approximately 70-80 % of the total fibers. This type of structure is designed to achieve uniformity in the mechanical strength of the composite materials. Steel fibers provide necessary strength and toughness to the sprayed concrete and prevent it from the brittle failure (El-Dieb, 2009). It is estimated that the applications of fiber jetting concrete in the coal mines will significantly improve the impact of the coal mine support systems.

## POTENTIAL USES OF COMPOSITE MATERIALS IN MINING INDUSTRY

Presently, China has become the most dynamic and rapidly growing country in the composite industry. In coming years, the tremendous economic progress of China will drive it to establish the global composite markets by bringing about developments in the composite industry. The US and other countries are also developing modern technologies to cope with the immediate challenges of the mining sector. The application of composites in the coal mines is essential for providing safety, energy conservation, industrial up-gradation and, efficient mining operations. Developing the composite materials according to the market demands is imperative. Thorough research is required to identify the specific needs of composite materials in the mining industry. However, the composite technology must be updated and upgraded on regular intervals to harvest the maximum results from the novel techniques. Innovations are essential for producing high-performance mining components. There is dire need to develop and utilize the composite materials in the mining industry to promote technological innovation, novel materials manufacturing, and socio-economic development. The scope of composite applications in the mining sector is pretty vast, and there are many unexplored aspects which need to be addressed.

Composite materials have a great potential to play a significant role in the mining industry. However, the real potential of the composite materials can only be realized if the composite industry is willing to take major risks by investing in the innovations and technological developments. Luckily, most of the modern technologies are available to the manufacturers and the mining industry, somehow, is benefiting from it. Eventually, it is imperative to develop an interactive bond between the mining industry and the composite industry. Bridging of both industries will result in the development of advanced technologies at cheaper rates. Moreover, this will ultimately boost the economy of the country.

## REFERENCES


1. (Markgraaf, J. (1995). Overview of new developments in composite materials for industrial and mining applications. Journal of the Southern African Institute of Mining and Metallurgy, 96(2), 55-65).
2. Waterman, N. A. (1991). Materials engineering for deep mining applications-an international view. Materials & Design, 12(1), 15-19.
3. Carter, F.C., & Paul, D.E. (1991). Material science and engineering. ASM
4. International.
5. Ashby, M. F., & Jones, D. R. (1995). Engineering materials 1. An introduction to their properties and applications. In Fuel and Energy Abstracts (Vol. 4, No. 36, p. 284).
6. Harris, B. (1991). A perspective view of composite materials development. Materials & Design, 12(5), 259-272.
7. Markgraaf, J. (1995). Overview of new developments in composite materials for industrial and mining applications. Journal of the

Southern African Institute of Mining and Metallurgy, 96(2), 55-65.

8. Crowson, P. (Ed.). (2016). Minerals Handbook 1996–97: Statistics & Analyses of the World's Minerals Industry. Springer.
9. De-Beer, Z.J. (1966). Applications for Composite Materials in the Mining and Metallurgy Industry, Present and Future. AERODYNE TECHNOLOGY.
10. Tenenbaum, D. J. (2009). TRASH OR TREASURE?: Putting coal combustion waste to work. Environmental health perspectives, 117(11), 491-497.
11. Park, H. C. (1992). Wear behavior of hybrid metal matrix composite materials. Scripta metallurgica et materialia, 27(4), 465-470.
12. Wang, Y. Q., Zhang, P. F., & Li, J. Z. (2013). The Application of Composite Materials in Coal Mine Production. In Advanced Materials Research (Vol. 756, pp. 49-53). Trans Tech Publications.
13. Jones, R. M. (1998). Mechanics of composite materials. CRC press.
14. Hanzawa, S., & Nakano, K. (2002). U.S. Patent No. 6,413,640. Washington, DC: U.S. Patent and Trademark Office.
15. Yungui, G., Zhengyou, W., & Mingcheng, L. (2001). APPLICATION OF GFRP PIPES IN COAL MINE [J]. Engineering Plastics Application, 8, 010.
16. Pickering, S. J. (2006). Recycling technologies for thermoset composite materials—current status. Composites Part A: Applied Science and Manufacturing, 37(8), 1206-1215.
17. Nanni, A., De Luca, A., & Zadeh, H. J. (2014). Reinforced concrete with FRP bars: mechanics and design. CRC Press.
18. MA, N. J., LI, Y. M., & JIE, A. Z. (2006). Molding technology of FRP bolt with a new structure. Fiber Reinforced Plastics, 1, 011.
19. Laney, P. (2002). Use of composite pipe materials in the transportation of natural gas. Idaho International Engineering and Environmental Laboratory, Bechtel BWXT Idaho, LLC.
20. Mohitpour, M., McManus, M., & Trefanenko, B. (2002, January). Trends in pipeline integrity inspection and rehabilitation techniques. In 2002 4th International Pipeline Conference (pp. 1493-1500). American Society of Mechanical Engineers.
21. Shamsuddoha, M., Islam, M. M., Aravinthan, T., Manalo, A., &

Lau, K. T. (2013). Effectiveness of using fibre-reinforced polymer composites for underwater steel pipeline repairs. Composite Structures, 100, 40-54.

22. Buckley, J. D., & Edie, D. D. (1993). Carbon-carbon materials and composites (Vol. 1254). William Andrew.
23. Pang, J., & Xu, D. (2007). Support technology of polypropylene fiber concrete shotcrete layer and its application to Guqiao coal mine. Yanshilixue Yu Gongcheng Xuebao/Chinese Journal of Rock Mechanics and Engineering, 26(5), 1073-1077.
24. El-Dieb, A. S. (2009). Mechanical, durability and microstructural characteristics of ultra-high-strength self-compacting concrete incorporating steel fibers. Materials & Design, 30(10), 4286-4292.

# CHAPTER 9

# USES OF COMPOSITE MATERIALS IN ELECTRONIC AND BIOSENSING APPLICATIONS

## BACKGROUND

Nanocomposites are advanced materials which have potential applications in electronic packaging. Nanocomposites are typically made up of nanofibers, and polymer matrices. Nanocomposites are extensively produced from carbon nanotubes as filler materials to enhance the electrical, mechanical and thermal properties. In electronic devices, heat dissipation is required along with high thermal conductivity. Here, the challenges and current development in the applications of nanotubes as filler material in the matrices of polymers are explored. The mixing of polymers and nanotubes results in what we call as Nanocomposites. Some of the most important issue related to the fabrication of Nanocomposites are:

- Interaction of carbon nanotube and polymer and interface nature
- Diffusion of carbon nanotube in the polymer matrix
- The arrangement of CNT's in a polymer host

This issue directly influences the thermal and electrical performance of the Nanocomposite. The recent Nanocomposites fabrication using carbon nanotubes as filler material and their potential application in packaging as a thermal interface are also described. Nanocomposite Fabrication has been a promising area of research for a long time because researchers have been dodged for finding the right combination of materials to develop the thermal interface material. This review gives the properties and history of carbon nanotubes are examined to know its importance as filler materials and the subsequent fabrication and design of Nanocomposites for thermal and electrical applications is discussed.

## DISCOVERY OF CARBON NANOTUBES (CNTS)

Nanocomposites are filler material of nanoscale dimensions embedded in a host of a polymer matrix. Various types of materials such as carbon fibers, boron nitride and silver particles have been embedded in polymer matrices over the years (Gonon & Boudefel, 2006; Lafdi et al., 2008; Ran & Gao, 2007). The discovery of single-walled and multiwalled carbon nanotubes in 1993 and 1991, respectively have enhanced the development and growth of carbon nanotubes related research (Iijima, 1991; Iijima & Ichihashi, 1993). The increase in Nanocomposites research has been enhanced by the first time incorporation of CNT's in a polymer matrix by Ahayan and team in 1994; they incorporated carbon nanotube in an epoxy matrix (Ajayan et al., 1994). In parallel with the fabrication and the relationship between properties and applications, it is necessary to review the history of characteristics of carbon nanotubes to have the gratitude of the force behind the development of Nanocomposites related research.

In 1889 Chambers and Hughes announced the production of carbon nanotubes followed by Schutzenberger who made a similar report in 1890. The development of TEM by Siemens which appeared commercially in 1939 ensured that structure of these fibers can now be clarified. Lukyanovich and Radushkevich investigated the production of hollow carbon fibers in 1992 as shown in Fig.80 (Grobert, 2007).

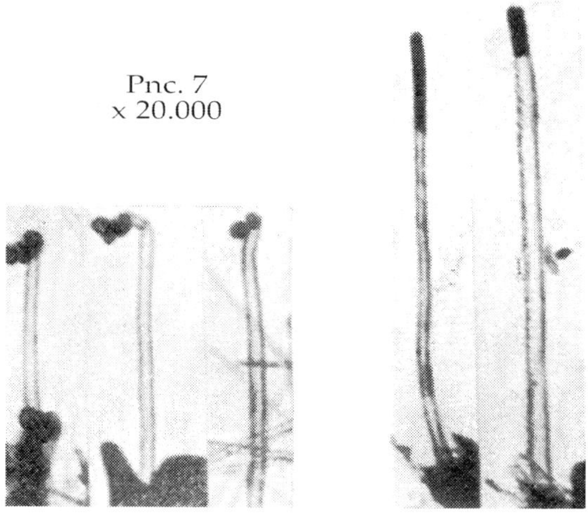


**Figure 80:** TEM image of hollow carbon fibers

These efforts marked the first success that led to the discovery of carbon nanotubes. Fig.81 shows structures which are similar to CNTs (Monthioux & Kuznetsov, 2006).

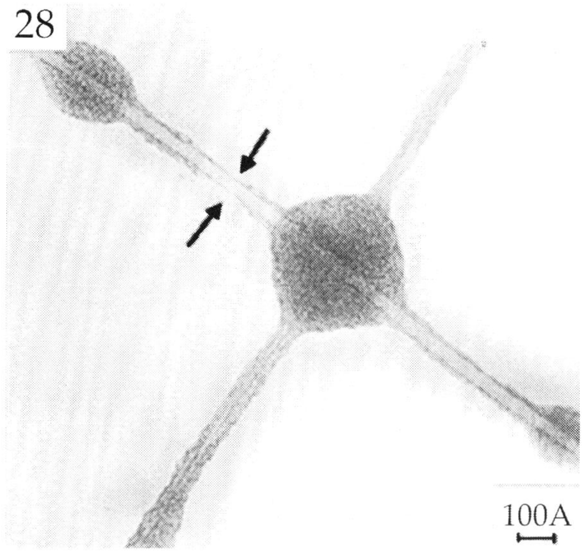


**Figure 81:** CNTs-like structures discovered by Monthioux & Kuznetsov (2006)

## Properties Structure and of Carbon Nanotubes

There are six electrons in carbon atoms which are arranged in such a way that two electrons out of six occupy the 1s orbital and other four electrons occupy $sp^2$ or $sp^3$ and also sp hybrid orbital. These orbitals which are hybrid orbitals are responsible for the bonding in graphite, nanotubes, fullerenes, and diamond. In graphite, each carbon atom has three outer shell electrons which occupy planar $sp^2$ orbital. This results in three in-plane $\sigma$ bonds and one out of plane $\pi$ bond, hence a network of planar hexagons as is shown in Fig.82.

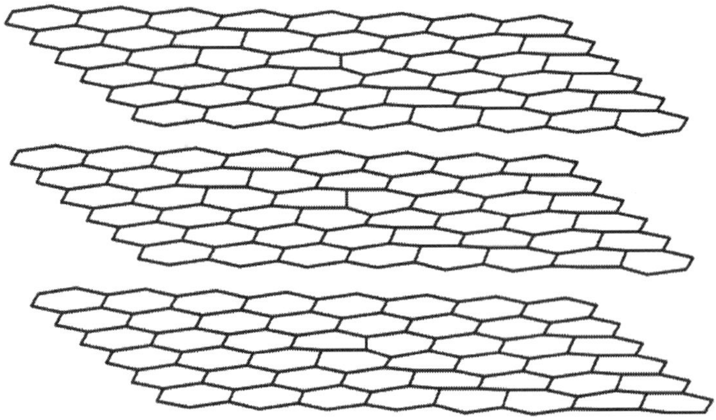

**Figure 82: Crystal structure of graphite layers**

The parallel layer in graphite network have a spacing of 0.34 nm and are bonded together by weak van der Walls forces. In $sp^2$ orbital, the $\sigma$ bond energy is 420 kcal/mol and bond length is 0.14 nm while in $sp^3$ orbital the bond energy is 360 kcal/mol and bond length is o.15 nm. Graphite is more electrically and thermally conductive because the $\pi$ bond which is out-of-plane is distributed over graphite plane. The formation of carbon nanotubes is done by rolling of graphite sheets which have $sp^2$ configuration (Han, 2005).

Graphene can be extracted from graphite by some methods such as chemical reduction of functionalised graphite oxide and nano-mechanical cleavage of graphite. Graphene is a monolayer, two-dimensional $sp^2$-bonded carbon atoms. It has been the area of interest for research since its development because of its exceptional transparent properties. Research is going on to use graphene as a conductive transparent electrode due to its high optical transparency. Graphene has high flexibility, chemical stability and mechanical strength as compared to traditional transparent electrodes such as ITO (tin-doped indium oxide). In Polymer Nanocomposites, graphene has also been used as a filler material, and it has been found that graphene could match with carbon nanotubes, it could be wise to use graphene to overcome the cost and toxicity associated with the carbon nanotubes. Flexible and transparent conductive materials are nowadays becoming famous in modern era providing critical components of video displays, plastic solar cells, and touch screens. In spite of the fact that ITO is brittle, indium is becoming expensive because of availability issues. Indium compounds have found to be relatively toxic.

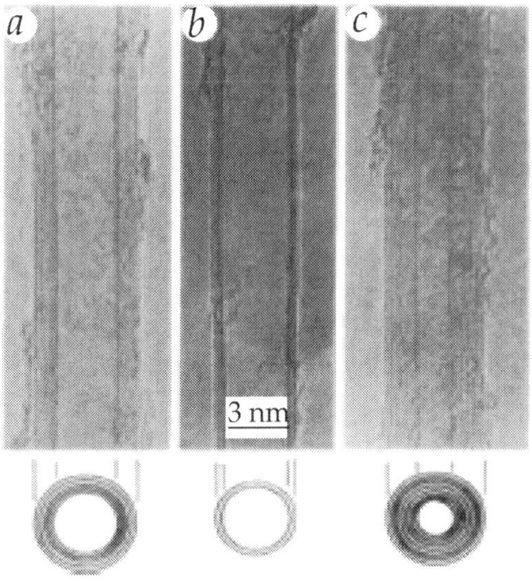


**Figure 83: TEM images of multi-walled carbon nanotubes (MWCNTs)**

Considering the current issues associated with modern material systems, researchers are now searching for something which would be environment-friendly, more stable, high conductivity, able to be processed in solution and highly transparent (Li et al., 2009). The conducting polymers such as poly (3, 4-ethylenedioxythiophene) (PEDOT) and hybrid material of graphene could be an alternative. The PEDOT-graphene hybrid material has excellent optical transparency, flexibility, solution processability and conductivity (Geng et al., 2009; Gilje et al., 2007; Xu et al., 2009). Although some structure similar to carbon nanotube had been observed earlier, none of these got as much importance as the publication, 'Helical microtubules of graphitic carbon' in 1991 by S. Iijima in the journal *Nature.* The results of Iijima high-resolution investigation of soot produced during the fabrication of fullerenes $C_{60}$ using the Huffman-Kraetschmer method (Kraetschmer et al., 1990). The CNT's were recognized as MWCNTs. Structures of CNTs are shown in Fig.83.

## Background and Current Trends

In 1993, D.S. Bethune also fabricated SWCNTs, but he used cobalt catalyst instead of iron as in the case of Ichihashi and Iijima (Bethune et al., 1993). After some years, Dai and his team claimed the existence of less number of double-walled carbon nanotubes along with single-walled nanotubes. They conducted experiments in 1996 by performing Disproportionation of Carbon Monoxide on alumina supported Mo particles at temperature of 1200 °C. The absence or existence of DWCNTs was linked with the nature of catalyst used by Hafner and his team (Dai et al., 1996). Fig.84 shows the structure of single-walled, double-walled and triple-walled carbon nanotubes.

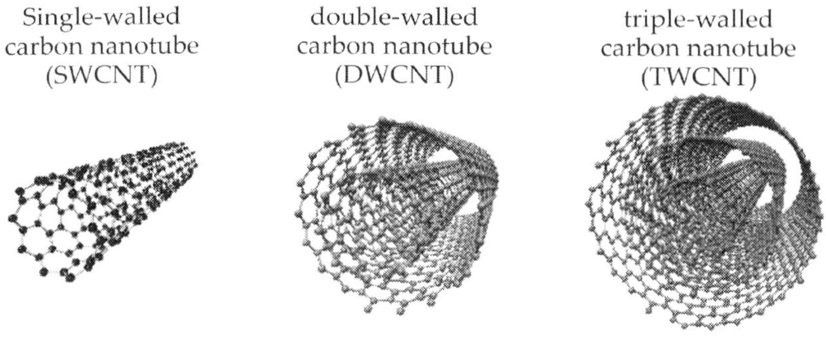


**Figure 84: Images of SWCNTs, DWCNTs, and TWCNTs**

In this time, carbon nanotubes are being fabricated by many methods such as laser-vaporization also called laser ablation, electrolysis, carbon arc discharge and chemical vapor deposition (Mylvaganam & Zhang, 2007). All the manufacturing methods tend to change the properties of the carbon nanotubes produced (Du et al., 2007. Typically, the critical properties of CNTs that have been reported in the literature are:

- The theoretical thermal conductivity of 6000 W/mK

- A measured value of 3000 W/mK
- Modulus of 1 Tera Pascal
- Tensile strength of 200 GPa
- The electron mobility of 105 cm2/Vs (Jaiswal & Menon, 2006; Coleman et al., 2006)

The micron-sized length and nanometer-sized diameter of the carbon nanotubes result in a high aspect ratio which is coupled with unique thermal, mechanical and electrical properties have resulted in scientists exploring how best to exploit them. A bulk carbon nanotube is fascinating for applications in electronic sensing devices and as filler material in polymer-based composites. Table 10 summarizes some of these properties (Kumari et al., 2009).

**Table 10:** Properties of carbon nanotubes

| **Carbon Nanotube Properties** | | |
|---|---|---|
| **Property SWCN** | **Ts** | **MWCNTs** |
| Aspect Ratio | 1000–10,000 | 1000–10,000 |
| Electrical Conductivity | $10^6$ S/m at 300K Individual | $>10^5$ S/m |
| Electron Mobility | $10^5$ cm$^2$/Vs | $10^5$ cm$^2$/Vs |
| Thermal Conductivity | 6600 W/mK Individual | 3000 W/mK individual |
| Mechanical | Young's Modulus: 1 Tera Pascal Tensile Strength: Up to 200 GPa | Young's Modulus: 1 Tera Pascal<br>Tensile Strength: Up to 200 GPa |

## NANOCOMPOSITES

Polymer reinforced with a network of fibers form a polymer matrix composite which is known for improving stiffness, tensile strength abrasion resistance, toughness, corrosion resistance and impact strength. The polymer materials can be used as without reinforcement, but their use is limited because of inferior mechanical properties. Polymer material usually has little strength, coupled with low impact strength.

To improve usability and durability, most of the polymer material are compounded with chemicals and other polymers. Compounding can be used to develop a polymer for specific applications using suitably selected additives (Bolgar et al., 2015).

## Polymer Nanocomposites

The polymer materials can be utilized as without reinforcement, but their use is limited because of inferior mechanical properties. Polymer material usually has little strength, coupled with low impact strength. Polymer Nanocomposites are now being used as an alternative to conventional polymers. In Polymer Nanocomposites, nanoscopic inorganic particles, which are usually 10-100 Ao in at least one dimension are distributed in a matrix of a polymer such as a polypyrrole, polythiophene and polyacetylene (Fischer, 2003). Because of their nanometer sizes of the filler, Nanocomposites exhibit improved properties when compared to the conventional polymers or their composites (Giannelis, 1998).

## Electrical and Thermal Conductivity of Nanocomposites

Various reports have been published on the thermal and electrical conductivity of the Nanocomposites at different loading using different polymer matrices and filler materials.

Table 11 shows the electrical percolation threshold of some of the Nanocomposites that have been fabricated. Table 12 summarizes the thermal conductivity measurements for selected Nanocomposites.

**Table 11:** Percolation threshold of Carbon Nanotube Nanocomposites (Jones et al., 2010)

| **Author** | **Sample** | **Percolation threshold [wt %]** |
|---|---|---|
| Shaffer & Windle (1999) | PVA + MWCNTs | 5 – 10 |
| Sandler *et al. (1998)* | Epoxy + CNTs | 0.0025 - 0.004 |
| Schadler *et al. (2003)* | Epoxy + aligned MWCNTs | 0.0025 |
| Allaoui *et al. (2002)* | Epoxy +CNTs | 0.5 – 1 |
| Andrews *et al. (2002)* | PS + MWCNTs | 0.25 vol % |
| Dufresne *et al. (2002)* | PS + MWCNTs | ≤3 |

| | | |
|---|---|---|
| Hu *et al. (2006)* | PET + MWCNTs | 0.9 |
| Barraza *et al. (2002)* | PS +SWCNTs | 8.5 |
| Seo *et al. (2005)* | PP + CNTs | 2 |
| McNally *et al. (2005)* | PE + MWCNTs | 7.5 |
| Zhang, *et al.(2006)* | HDPE + SWCNTs | 4 |
| Seo *et al. (2004)* | PP + CNTs | 1 - 2 |
| Kymakis *et al. (2002)* | P3OT + SWCNTs | 11 |
| Coleman *et al. (1998)* | PmPV + CNTs | 8.5 |
| Meincke *et al. (2004)* | PA6 + CNTs | 4 – 6 |
| Tchmutin *et al. (2003)* | PP + SWCNTs | 4.5 vol % |

Table 11 shows that smallest amount of carbon nanotube required to manufacture percolating network that allows the electrical conductivity in the Nanocomposite is significantly low. Apart from the type of carbon nanotube and type of polymer matrix, the most important factor that changes the percolation threshold is an arrangement of the carbon nanotubes. The minimum percolation threshold in Table 12 of weight % 0.0025 was observed for an aligned MWCNT epoxy composite.

**Table 12:** Thermal conductivity of some selected Nanocomposites

| **Authors** | **Samples** | **Thermal Conductivity [W/mK]** |
|---|---|---|
| Huang *et al. (2005)* | Silicone elastomer + 0.4wt % MWNTs | Aligned: 0.65, Dispersed: 0.31 |
| Liu *et al.* (2004) | Polydimethylsiloxane rubber + 2wt % MWCNTs | 0.190 |
| Haggenmueller *et al. (2007)* | HDPE + 30wt % SWCNTs | 3.5 |
| Song *at al. (2006)* | Silicone elastomer + 1.6wt % MWCNTs | 1.59 |
| Bryning *et al.(2005)* | Epoxy+1wt % SWCNTs | 80% enhancement |
| Peters *et al. (2008)* | PS + 30wt % MWCNTs | >0.6 |
| Liu *et al. (2004)* | Silicone elastomer + 3.8wt % SWCNTs+MWCNTs | 65% enhancement |
| Fukushima *et al.(2006)* | Nylon 6 + 20 vol % xGnP (graphite) | >4 |

| | | |
|---|---|---|
| Du *et al. (2006)* | Epoxy + SWCNTs | 0.61 |
| Biercuk *et al.* (2002) | Epoxy + SWCNTs | 125% enhancement |

The data of thermal conductivity in Table 12 indicates that carbon nanotubes loaded composites have improved thermal conductivity as compared to the polymer which normally has thermal conductivity if about 0.2 W/mK. (Guthy et al., 2007). The data also indicated a remarkable increase in conductivity when graphite nanofiber are induced as filler. A thermal conductivity value of greater than 4 W/mK was observed at a loading of 20 vol %. Even though there are signs of improvement in electrical and thermal conductivity, the value is still significantly below the expected by theory.

**Table 13**: Electrical Conductivity of Polycarbonate PC/MWCNTs Nanocomposites

| Sample Percolation | | **Threshold (wt %)** | **Electrical Conductivity (S/cm)** |
|---|---|---|---|
| PC/dispersed MWCNTs | | ~0.1 | $8.14 \times 10^{-5}$ |
| PC/ dispersed MWCNTs | | ~2 | $2.87 \times 10^{-5}$ |
| PC/dispersed MWCNTs | | ~5 | $5.69 \times 10^{-6}$ |
| PC/dispersed MWCNTs | | ~5 | $3.65 \times 10^{-4}$ |
| PC/dispersed MWCNTs | | ~5 | $2.08 \times 10^{-3}$ |

Our latest work on the manufacturing and characterization of Nanocomposite materials for thermal and electrical conductivity has revealed the dependence of properties on the post-production treatment, the type polymer matrix and the dispersion method (Chiguma, 2009). Table 13 and Table 14 indicate some example of polymer based Nanocomposites that were manufactured in our group. The data consists of thermal conductivity and percolation threshold. In Table 14, the four samples were prepared from polycarbonate and the same type of MWCNTs that had given different post-production treatments. The percolation threshold for a single sample is around two weight % while that of the others start at around 5wt %. Fascinatingly, the electrical conductivities change as well, indicating the effect of post-production treatment on electrical conductivity and percolation threshold of Nanocomposites. The last sample in Table 13 was prepared from polycarbonate and a different type of MWCNT that was fabricated by chemical vapor deposition. The electrical conductivity was found to be around $8.14 \times 10^{-5}$ S/cm, and percolation threshold was around 0.1wt %.

**Table 14:** Thermal Conductivity of Epoxy/MWCNTs Nanocomposites

| Sample | Thermal Conductivity (W/mK) |
|---|---|
| Epoxy | 0.20 |
| Epoxy/0.1wt % MWCNTs (dispersed) | 1.85 |
| Epoxy/ 0.1 wt% MWCNTs (dispersed) | 1.91 |
| Epoxy/0.1wt % MWCNTs (dispersed) | 1.55 |

The results of various post-production treatments on the thermal conductivity of Nanocomposites can be observed in Table 5. The MWCNTs were produced by the combustion process except that they were given different treatment after production. At the same percentage loading of 0.1wt %, the Nanocomposites had different thermal conductivity values.

## Enhancement in Properties

Polymer Nanocomposite has shown evidently the property advantage that nano-sized additives can give in comparison to both traditional base polymer and filler counterparts. Properties which have been revealed to undergo improvement are:

- Mechanical properties e.g. modulus, dimensional stability, and strength decreased flammability and Improved heat and solvent resistance
- Heat distortion temperature and thermal stability
- Optical clarity in comparison to traditionally filled polymers
- Decreased permeability to hydrocarbons, water, and gases
- Reduced smoke emissions and high flame resistance
- Electrical conductivity
- Chemical resistance
- Surface appearance

# APPLICATIONS

Nanocomposites and Nanotechnology are deeply rooted in the design of advanced devices for optoelectronic and electronic applications. The dimensional scale for electronic devices has now come in the nano range. The use of polymer-based Nanocomposites in these areas is quite varied involving many applications and have been considered for their use in various applications.

## Typical Applications of Nanocomposites

They are employed in smart windows and memory devices, chemical sensors, electrocatalysis, batteries and electroluminescent devices. Working principle of smart windows is shown in Fig.85.

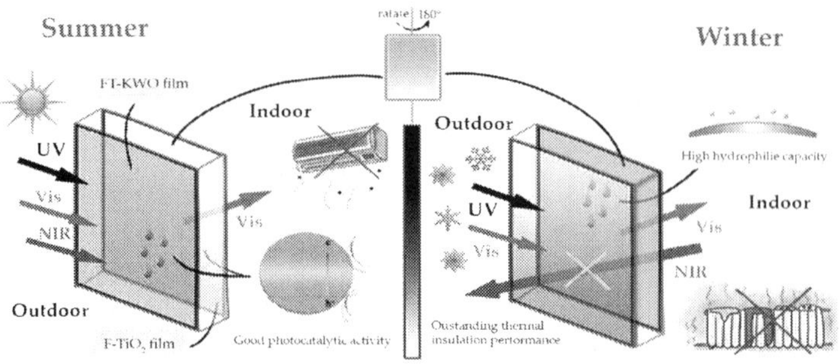


**Figure 85:** Working principle of Nanocomposite based smart windows

Other potential applications of Nanocomposites include light emitting diodes (LEDs), photodiodes photovoltaic (PV) cells, super-capacitors, field effect transistors and printable conductors (Tyagi, 2014).

**Figure 86:** Light emitting diodes (LEDs) produced from Nanocomposites

It is also an interesting topic to study the electrical conductivity of carbon nanotubes in insulating polymers. The potential applications are transparent conductive coatings, electromechanical actuators, and electrostatic dissipation, including electromagnetic interference shielding, super capacitors and various electrode applications (Wang et al., 2003).

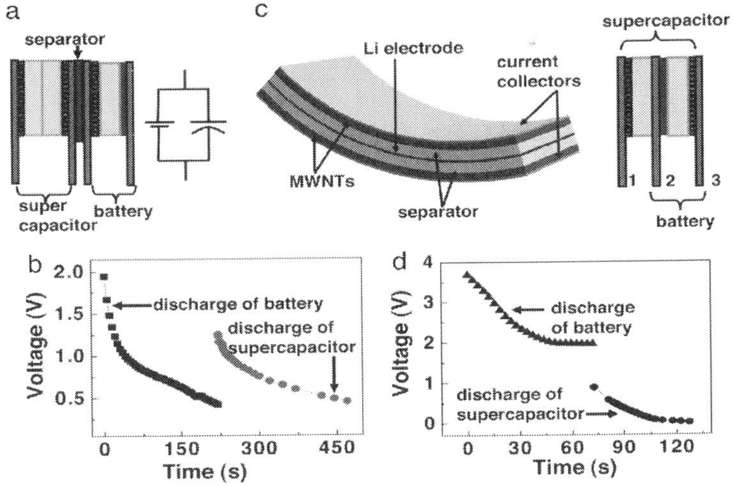


**Figure 87**: (a) Configuration, (b) schematic of working principle and (c) structure of a typical Nanocomposite super-capacitor

Conjugated polymers with different nanoscale filler materials have been studied for sensor applications including biosensors, chemical sensors and gas sensors. The Nano-fillers employed may include carbon nanotubes, metal oxide nanowires, nanoscale gold, copper, nickel, silver, palladium and platinum particles (Sandler et al., 2003).

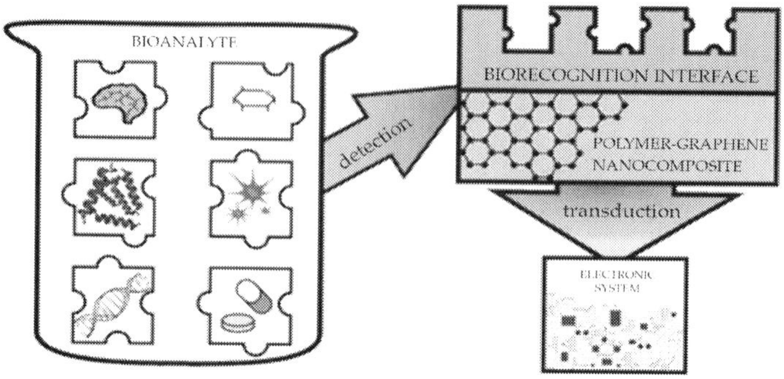


**Figure 88:** Working principle of a Nanocomposite biosensor

The interest of graphene is also in parallel to that of carbon nanotubes in electronic applications. Graphene can be used to replace silicon when used in sheet form as Moore's law limits are achieved (Van Noorden, 2006).

Cheap, big and flexible solar panels can be made by using polymer-based solar cells. But the only disadvantages of these panels would be low efficiency as compared to commercial solar cells (Krebs, 2009). Cadmium chalcogenides nanoparticles have been used to make solar cells in a polymer matrix. Hybrid composite coatings are also being fabricated for developing thin film solar cells. The efficiency of hybrid solar cells is comparatively higher than the conventional thin film solar cells. Moreover, composite coatings are also applied in satellite electronics.

**Figure 89:** Flexible solar panels prepared from Nanocomposites

Polymer semiconductor Nanocomposite materials have promising application for generation ofhybrid materials which have various potential applications such as superconductor devices, photoconductors, visual displays, gas sensors, mechanics, electrical devices and photovoltaics (Kothurkar, 2004).

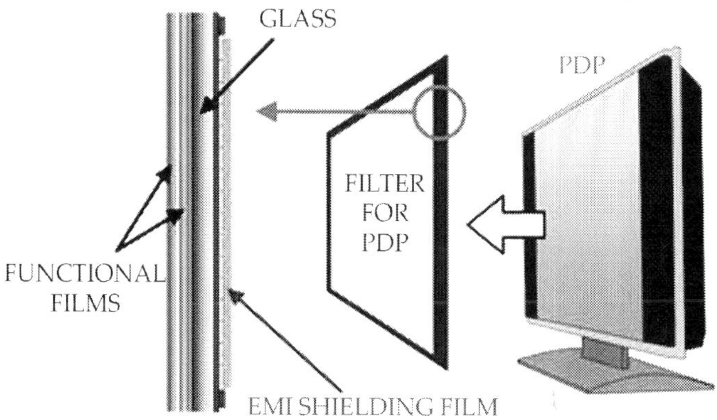


**Figure 90:** Composite shields embedded in display screens for EMI and RFI protection

## Nanocomposites for Sensor Applications

Nanomaterials have a variety of applications in electrochemical biosensing thereby improving the performance of products and opening new fields in their uses. Nanotubes, nanoparticles, and nanowires have made an impact in on the areas of biosensors, ranging from Gen electronic sensors to glucose enzyme electrodes. Due to some characteristic features of conducting polymer Nanocomposites like light weight, glucose enzyme electrodes, chemical specificities, have large surface area, sound processing and scalable productions and low cost, they are used in applications like biological and chemical sensors and nanoelectronic devices (Sree et al., 2002).

Manufacturing of biosensors using nanomaterials allows the usage of a variety of new signal transduction technologies (Jianrong et al., 2004). In molecular sensor and electronics, CPs have been used as a potential system for control of enzymes (Wang, 2005; Takashima & Kaneto, 2004). In this system, direct transfer of electrons is to and from enzymes.

The setup of enzymes in CP films gives a controlled method of localizing active molecules indefinite area on the electrodes. In bioanalytical sciences the use of conducting polymers is of high interest because they are biocompatible and are used efficiently in vivo biosensors, for metabolites in biological fluids or monitoring of drugs (Geeta et al., 2006).

Using interfacial polymerization, polypyrrole Nanofilms were prepared using sulfate as a dopant to make insertion of different functional groups to pyrrole film and give various applications in fabricating biological and chemical sensors (McQuade et al., 2000).

Nanoparticles provide a reasonable solution to the problems related to optimization of immobilized enzymes: maximum surface area per unit mass, high active enzyme loading and minimum diffusion limitations (Ramanavicius et al., 2005).

Nowadays, protocols based on nanoparticles are being studied for the detection of proteins. The property linked with the nanotubes and nanowires, which make it easy for us to modify them for biological recognition elements, gives high selectivity to these devices. Electrochemical sensors based on Nanomaterials are expected to develop a significant influence upon the security surveillance, environmental monitoring, for ensuring our food safety and clinical diagnosis.

### *Gas Sensors*

Thin films or the blends of conducting polymers as sensors for airborne volatiles ($NH_3$, alcohols, CO, $NO_2$) has been utilized extensively. Polythiophene-based sensors have also been used for the detection of ppb of hydrazine gases. Ultra-thin films of polyaniline–$SnO_2/TiO_2$ Nanocomposite has been developed for the CO gas sensing (Ram et al., 2005). By using MWNTCOOH/PPy-modified glassy carbon electrode a novel sensitive electrochemical biosensor for monitoring DNA hybridization was prepared. The detection limit of this biosensor is $2.3 \times 10^{-14}$ $mol^{-1}$ and range is found to be $6.9 \times 10^{-14}$–$8.6 \times 10^{-13}$ $mol^{-1}$ (Cheng et al., 2005).

### *Biosensors*

Conducting Polymer Nanocomposite sensor for glucose is useful for in vivo detection (Forzani et al., 2004). Junctions were formed by bridging a pair of nano-electrodes separated by a gap of 20-60 nm with PANI/$GO_x$. The fabrication of novel electrochemical DNA biosensor based on methylene blue and polyaniline nanowires for DNA hybridization has been presented (Zhu et al., 2005). The sensitivity of this method was very fascinating, and limit of detection reaches $1.0 \times 10^{-12}$ mol $l^{-1}$. It has been shown that conducting nano and micro containers can be made by electrochemical polymerization of monomers (Bajpal et al., 2006). These containers are fascinating for a large number of applications from sensors to release of drugs control. There are many such kinds of reports that produce the use of conducting polymers for developing nanosensors for DNA detection (Cai et al., 2003; Hahm & Lierber, 2004). Recently, for the detection of "bisphenol A" an impedimetric immunogen was developed through covalent bond formation.

For label-free affinity sensing the conducting polymer nanowire sensors have shown to be fascinating. For example, the monitoring of the nanomolar concentration of biotin at an avidin-embedded polypyrrole has been demonstrated by Ramanathan et al. (2005). In other research, the use of polypyrrole nanowire electrodes due to their Amperometric response towards nitrate ions is shown. The detection limit and sensitivity were found to be and $1.52 \times 10^{-6}$ M and 336.28 mA $M^{-1}$ $cm^{-2}$ respectively. Another selective and highly sensitive nitrate ions sensor has been shown by using doping approach on PPy nanowires.

The possibility of manufacturing polyaniline nanowires and single polypyrrole and their application as DNA sensors (1 nm) were also examined by Im et al. (2006). Such as glucose sensor has been fabricated and characterized, based on glucose oxidase onto a low-noise carbon fiber nano-electrode and co-electrodeposition of redox polymer poly (vinyl imidazole).

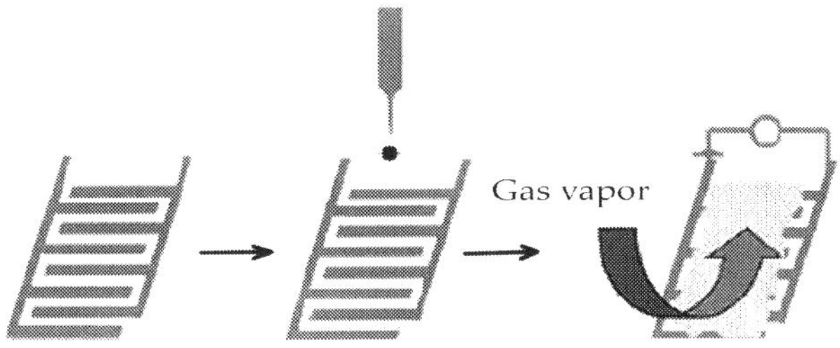


Figure 91: Schematic of a typical sensor experimentation

### *CNTS based Sensors*

This Nano sensor suggests a highly quick and sensitive response to glucose at a working potential of 0.22 V, with a detection limit of 0.004 mM and a linear range of 0.01–15 mM. The use of conducting polymer substrates and the amplification of semiconductor nanoparticles can be united to develop a novel DNA sensor (Peng et al., 2006). Lately, conducting polymer/CNTs, composites have established substantial interest because the combination of CNTs and conducting polymers can lead to new composite materials having the properties of each component with a synergistic effect. The subtle electronic resources recommend that CNT has the capability to encourage electron transfer reaction when used as an electrode. CNT/polymer composites have been used for control of enzymes and metalloproteinases by either covalent binding or physical adsorption.

Polyaniline and Polypyrrole can be employed for production of CNT/PANI and CNT/PPy Nanocomposite electrodes due to the ease in the making through copolymerization by an electrochemical or chemical approach, and the resulting nanocomposites show high stability and conductivity (Baibarac et al., 2003). PANI/CNTs composite modified electrode developed by Galvano static electro polymerization of aniline exhibits improved electrolytic behavior to the reduction of nitrite and facilitated the recognition of nitrite at a potential of 0.0 voltage. A sensitivity of 719.0 mA $M^{-1}$ $cm^{-2}$, the detection limit of 1.0 M and linear range from $5.0 \times 10^{-6}$ to $1.5 \times 10^{-2}$ M for the detection of sodium nitrite has been observed at the PANI/MWNTs-modified electrode.

The excellent electrical properties and surface area provided by the CNT's is a fascinating feature in the development of biological/chemical biosensors. Mostly, gas sensing is performed by CNTs by measuring the change in electrical properties of CNTs induced by or the mass change due to physical adsorption of gas molecules or the change transfer with gas molecules. Glucose oxidase which contains aligned carbon nanotube coaxial nanowire electrodes was made and utilized as a new glucose sensor. In ultra-thin polymer layer, the 3-D structure of CNTs provides an excellent model for enzymes loading, leading to a response of glucose of 10-20 times greater than that of flat electrode. A hydroquinone sensor was developed by Kong et al. on the synergistic effect of conducting poly (N-acetylaniline) polymer and MWCNT, with reproducibility and stability of electrode and sensitive detection. Over-oxidation of PPy–MWCNT composite film was studied by Tu et al. (2007) in alkaline and neutral solutions by the impedance of electrochemical quartz crystal. This OPPy/CNT/NaOH/Au electrode was used for sensing dopamine with a detection limit down to 1.7 $nmol^{-1}$. This efficient biosensor format shows a greater signal to background ratios and smaller response times.

### Sensors Produced from Hybrid Nanocomposites

Nanocomposite materials containing carbon black, MWNT and $TiO_2$ nanoparticles were deposited in thin films to examine the impedance features for biosensing applications (Carrara et al., 2005). These hybrid composite structures were deposited by electrodeposition and various other techniques to investigate their functional and electronic properties.

## FUTURE PROSPECTS OF NANOCOMPOSITES IN ELECTRONIC INDUSTRY

If we look at the 21st century, nanoscience will be one the promising fields that will contribute to the high level technological and scientific development. Hybrid inorganic-(bio)-organic materials will be considered necessary in the elaboration of the advanced functional materials. Research in hybrid inorganic-organic materials is being supported by the growing interests of materials scientists, biologists, physicists and chemists to fully exploit the chance for making smart materials from best of three areas: organic, inorganic and biological. The advancement in nanostructured conducting polymer materials and Polymer Nanocomposites have a significant influence on biological research.

Considerable advances in the manufacturing of Nano-biosensors are regularly made. In this review, we shortly described the methods which provide various synthetic routes with benefits and shortcomings therein to fabricate the nanostructured conducting polymers. The study also shows the importance of nano-structured conducting polymers in developing fields of nanosensors. A comprehensive and detailed analysis has been carried out on the modern research advancement made in the fabrication of nanostructured polymer Nanocomposites and conducting polymers based biosensors. As the surface nanostructure become more complex and demanding, more methods for the construction of nanostructured materials will be required. These methods will use new approaches to conducting polymers and their applications in biomedical research. Increasing interest in and practical use of nanotechnology, especially, in conducting polymers have led the researchers to the rapid development of nanosensors with improved functionality and processability over previously developed biosensor.

## REFERENCES

1. Ajayan, P. M., Stephan, O., Colliex, C., & Trauth, D. (1994). Aligned carbon nanotube arrays formed by cutting a polymer resin-nanotube composite. Science-AAAS-Weekly Paper Edition, 265(5176), 1212-1214.
2. Allaoui, A., Bai, S., Cheng, H. M., & Bai, J. B. (2002). Mechanical and electrical properties of an MWNT/epoxy composite. Composites science and technology, 62(15), 1993-1998.
3. Andrews, R., Jacques, D., Minot, M., & Rantell, T. (2002). Fabrication of carbon multiwall nanotube/polymer composites by shear mixing. Macromolecular Materials and Engineering, 287(6), 395-403.
4. Baibarac, M., Baltog, I., Lefrant, S., Mevellec, J. Y., & Chauvet, O. (2003). Polyaniline and carbon nanotubes based composites containing whole units and fragments of nanotubes. Chemistry of Materials, 15(21), 4149-4156.
5. Bajpai, V., He, P., Goettler, L., Dong, J. H., & Dai, L. (2006). Controlled syntheses of conducting polymer micro and nanostructures for potential applications. Synthetic Metals, 156(5), 466-469.
6. Barraza, H. J., Pompeo, F., O'Rea, E. A., & Resasco, D. E. (2002). SWNT-filled thermoplastic and elastomeric composites prepared by mini-emulsion polymerization. Nano Letters, 2(8), 797-802.
7. Bethune, D. S., Klang, C. H., De Vries, M. S., Gorman, G., Savoy, R., Vazquez, J., & Beyers, R. (1993). Cobalt-catalysed growth of carbon nanotubes with single-atomic-layer walls. Nature, 363(6430), 605-607.
8. Biercuk, M. J., Llaguno, M. C., Radosavljevic, M., Hyun, J. K.,

Johnson, A. T., & Fischer, J. E. (2002). Carbon nanotube composites for thermal management. Applied physics letters, 80(15), 2767-2769.

9. Bolgar, M., Hubball, J., Groeger, J., & Meronek, S. (2015). Handbook for the chemical analysis of plastic and polymer additives. CRC Press.
10. Cai, H., Xu, Y., He, P. G., & Fang, Y. Z. (2003). Indicator free DNA hybridization detection by impedance measurement based on the DNA-doped conducting polymer film formed on the carbon nanotube modified electrode. Electroanalysis, 15(23-24), 1864-1870.
11. Carrara, S., Bavastrello, V., Ricci, D., Stura, E., & Nicolini, C. (2005). Improved nanocomposite materials for biosensor applications investigated by electrochemical impedance spectroscopy. Sensors and Actuators B: Chemical, 109(2), 221-226.
12. Cheng, G., Zhao, J., Tu, Y., He, P., & Fang, Y. (2005). A sensitive DNA electrochemical biosensor based on magnetite with a glassy carbon electrode modified by multi-walled carbon nanotubes in polypyrrole. Analytica Chimica Acta, 533(1), 11-16.
13. Chiguma, J. (2009). Conducting polymer nanocomposites loaded with nanotubes and fibers for electrical and thermal applications. The state University of New York at Binghamton.
14. Coleman, J. N., Curran, S., Dalton, A. B., Davey, A. P., McCarthy, B., Blau, W., & Barklie, R. C. (1998). Percolation-dominated conductivity in a conjugated-polymer-carbon-nanotube composite. Physical Review B, 58(12), R7492.
15. Coleman, J. N., Khan, U., Blau, W. J., & Gun'ko, Y. K. (2006). Small but strong: a review of the mechanical properties of carbon nanotube–polymer composites. Carbon, 44(9), 1624-1652.
16. Dai, H., Rinzler, A. G., Nikolaev, P., Thess, A., Colbert, D. T., & Smalley, R. E. (1996). Single-wall nanotubes produced by metal-catalyzed disproportionation of carbon monoxide. Chemical Physics Letters, 260(3-4), 471-475.
17. Du, F., Guthy, C., Kashiwagi, T., Fischer, J. E., & Winey, K. I. (2006). An infiltration method for preparing single-wall nanotube/epoxy composites with improved thermal conductivity. Journal of

Polymer Science Part B: Polymer Physics, 44(10), 1513-1519.

18. Du, J. H., Bai, J., & Cheng, H. M. (2007). The present status and key problems of carbon nanotube based polymer composites. Express Polym Lett, 1(5), 253-273.
19. Dufresne, A., Paillet, M., Putaux, J. L., Canet, R., Carmona, F., Delhaes, P., & Cui, S. (2002). Processing and characterization of carbon nanotube/poly (styrene-co-butyl acrylate) nanocomposites. Journal of materials science, 37(18), 3915-3923.
20. Fischer, H. (2003). Polymer nanocomposites: from fundamental research to specific applications. Materials Science and Engineering: C, 23(6), 763-772.
21. Forzani, E. S., Zhang, H., Nagahara, L. A., Amlani, I., Tsui, R., & Tao, N. (2004). A conducting polymer nanojunction sensor for glucose detection. Nano Letters, 4(9), 1785-1788.
22. Fukushima, H., Drzal, L., Rook, B., & Rich, M. (2006). The thermal conductivity of exfoliated graphite nanocomposites. Journal of thermal analysis and calorimetry, 85(1), 235-238.
23. Geetha, S., Rao, C. R., Vijayan, M., & Trivedi, D. C. (2006). Biosensing and Drug delivery by polypyrrole. Analytica Chimica Acta, 568(1), 119-125.
24. Geng, Y., Wang, S. J., & Kim, J. K. (2009). Preparation of graphite nanoplatelets and graphene sheets. Journal of colloid and interface science, 336(2), 592-598.
25. Giannelis, E. P. (1998). Polymer-layered silicate nanocomposites: synthesis, properties, and applications.
26. Gilje, S., Han, S., Wang, M., Wang, K. L., & Kaner, R. B. (2007). A chemical route to graphene for device applications. Nano Letters, 7(11), 3394-3398.
27. Gonon, P., & Boudefel, A. (2006). Electrical properties of epoxy/silver nanocomposites. Journal of applied physics, 99(2), 024308.
28. Grobert, N. (2007). Carbon nanotubes–becoming clean. Materials today, 10(1), 28-35.
29. Guthy, C., Du, F., Brand, S., Winey, K. I., & Fischer, J. E. (2007). The thermal conductivity of single-walled carbon nanotube/PMMA nanocomposites. Journal of heat transfer, 129(8), 1096-1099.

30. Haggenmueller, R., Guthy, C., Lukes, J. R., Fischer, J. E., & Winey, K. I. (2007). Single-wall carbon nanotube/polyethylene nanocomposites: thermal and electrical conductivity. Macromolecules, 40(7), 2417-2421.
31. Hahm, J. I., & Lieber, C. M. (2004). Direct ultrasensitive electrical detection of DNA and DNA sequence variations using nanowire nanosensors. Nano Letters, 4(1), 51-54.
32. Han, J. (2005). Carbon Nanotubes Science and Applications. Meyyappan, M., Ed.; CRC Press LLC: Boca Raton, Florida, USA, 2.
33. Hu, G., Zhao, C., Zhang, S., Yang, M., & Wang, Z. (2006). Low percolation thresholds of electrical conductivity and rheology in poly (ethylene terephthalate) through the networks of multi-walled carbon nanotubes. Polymer, 47(1), 480-488.
34. Huang, H., Liu, C. H., Wu, Y., & Fan, S. H. O. U. S. H. A. N. (2005). Aligned carbon nanotube composite films for thermal management. Advanced Materials, 17(13), 1652-1656.
35. Iijima, S. (1991). Helical microtubules of graphitic carbon. Nature, 354(6348), 56.
36. Iijima, S., & Ichihashi, T. (1993). Single-shell carbon nanotubes of 1-nm diameter. Nature, 363(6430), 603-605.
37. Im, Y., Vasquez, R. P., Lee, C., Myung, N., Penner, R., & Yun, M. (2006). Single metal and conducting polymer nanowire sensors for chemical and DNA detections. In Journal of Physics: Conference Series (Vol. 38, No. 1, p. 61). IOP Publishing.
38. Jaiswal, M., & Menon, R. (2006). Polymer electronic materials: a review of charge transport. Polymer International, 55(12), 1371-1384.
39. Jianrong, C., Yuqing, M., Nongyue, H., Xiaohua, W., & Sijiao, L. (2004). Nanotechnology and biosensors. Biotechnology Advances, 22(7), 505-518.
40. Jones, W. E., Chiguma, J., Johnson, E., Pachamuthu, A., & Santos, D. (2010). Electrically and thermally conducting nanocomposites, for electronic applications. Materials, 3(2), 1478-1496.
41. Krätschmer, W., Lamb, L. D., Fostiropoulos, K., & Huffman, D. R.

(1990). Solid C60: a new form of carbon. Nature, 347, 27.

42. Krebs, F. C. (2009). Fabrication and processing of polymer solar cells: a review of printing and coating techniques. Solar energy materials and solar cells, 93(4), 394-412.
43. Kothurkar, N. K. (2004). Solid state, transparent, cadmium sulfide-polymer nanocomposites (Doctoral dissertation, University of Florida).
44. Kumari, L., Zhang, T., Du, G. H., Li, W. Z., Wang, Q. W., Datye, A., & Wu, K. H. (2009). Synthesis, microstructure and electrical conductivity of carbon nanotube–alumina nanocomposites. Ceramics International, 35(5), 1775-1781.
45. Kymakis, E., Alexandou, I., & Amaratunga, G. A. J. (2002). Single-walled carbon nanotube–polymer composites: electrical, optical and structural investigation. Synthetic Metals, 127(1), 59-62.
46. Lafdi, K., Fox, W., Matzek, M., & Yildiz, E. (2008). Effect of carbon nanofiber-matrix adhesion on polymeric nanocomposite properties: part II. Journal of Nanomaterials, 2008, 5.
47. Li, X., Zhu, Y., Cai, W., Borysiak, M., Han, B., Chen, D., & Ruoff, R. S. (2009). Transfer of large-area graphene films for high-performance transparent conductive electrodes. Nano Letters, 9(12), 4359-4363.
48. Liu, C. H., Huang, H., Wu, Y., & Fan, S. S. (2004). Thermal conductivity improvement of silicone elastomer with carbon nanotube loading. Applied Physics Letters, 84(21), 4248-4250.
49. McNally, T., Pötschke, P., Halley, P., Murphy, M., Martin, D., Bell, S. E., & Quinn, J. P. (2005). Polyethylene multiwalled carbon nanotube composites. Polymer, 46(19), 8222-8232.
50. McQuade, D. T., Pullen, A. E., & Swager, T. M. (2000). Conjugated polymer-based chemical sensors. Chemical Reviews, 100(7), 2537-2574.
51. Meincke, O., Kaempfer, D., Weickmann, H., Friedrich, C., Vathauer, M., & Warth, H. (2004). Mechanical properties and electrical conductivity of carbon nanotube filled polyamide-6 and its blends with acrylonitrile/butadiene/styrene. Polymer, 45(3), 739-748.
52. Monthioux, M., & Kuznetsov, V. L. (2006). Who should be given

credit for the discovery of carbon nanotubes? Carbon, 44(9), 1621-1623.

53. Mylvaganam, K., & Zhang, L. C. (2007). Fabrication and application of polymer composites comprising carbon nanotubes. Recent Patents on Nanotechnology, 1(1), 59-65.
54. Peters, J. E., Papavassiliou, D. V., & Grady, B. P. (2008). Unique thermal conductivity behavior of single-walled carbon nanotube–polystyrene composites. Macromolecules, 41(20), 7274-7277.
55. Ram, M. K., Yavuz, Ö., Lahsangah, V., & Aldissi, M. (2005). CO gas sensing from ultrathin nano-composite conducting polymer film. Sensors and Actuators B: Chemical, 106(2), 750-757.
56. Ramanathan, K., Bangar, M. A., Yun, M., Chen, W., Myung, N. V., & Mulchandani, A. (2005). Bioaffinity Sensing using biologically functionalized conducting polymer nanowire. Journal of the American Chemical Society, 127(2), 496-497.
57. Ramanavičius, A., Kaušaitė, A., & Ramanavičienė, A. (2005). Polypyrrole-coated glucose oxidase nanoparticles for biosensor design. Sensors and Actuators B: Chemical, 111, 532-539.
58. Ran, S., & Gao, L. (2007). Electroconductive Al2O3–NbN Composites. Journal of the American Ceramic Society, 90(8), 2626-2629.
59. Sandler, J. K. W., Kirk, J. E., Kinloch, I. A., Shaffer, M. S. P., & Windle, A. H. (2003). Ultra-low electrical percolation threshold in carbon-nanotube-epoxy composites. Polymer, 44(19), 5893-5899.
60. Schadler, L. S., Giannaris, S. C., & Ajayan, P. M. (1998). Load transfer in carbon nanotube epoxy composites. Applied physics letters, 73(26), 3842-3844.
61. Seo, M. K., & Park, S. J. (2004). Electrical resistivity and rheological behaviors of carbon nanotubes filled polypropylene composites. Chemical Physics Letters, 395(1), 44-48.
62. Seo, M. K., Lee, J. R., & Park, S. J. (2005). Crystallization kinetics and interfacial behaviors of polypropylene composites reinforced with multi-walled carbon nanotubes. Materials Science and Engineering: A, 404(1), 79-84.
63. Shaffer, M. S., & Windle, A. H. (1999). Fabrication and

characterization of carbon nanotube/poly (vinyl alcohol) composites. Advanced Materials, 11(11), 937-941.

64. Song, P. C., Liu, C. H., & Fan, S. S. (2006). Improving the thermal conductivity of nanocomposites, by increasing the length efficiency of loading carbon nanotubes. Applied Physics Letters, 88(15), 153111.
65. Sree, U., Yamamoto, Y., Deore, B., Shiigi, H., & Nagaoka, T. (2002). Characterisation of polypyrrole nano-films for membrane-based sensors. Synthetic Metals, 131(1), 161-165.
66. Takashima, W., & Kaneto, K. (2004). Amperometric phenol biosensor based on covalent immobilization of tyrosinase onto an electrochemically prepared novel copolymer poly (N-3-aminopropyl pyrrole-co-pyrrole) film. Sensors and Actuators B: Chemical, 102(2), 271-277.
67. Tchmutin, I. A., Ponomarenko, A. T., Krinichnaya, E. P., Kozub, G. I., & Efimov, O. N. (2003). Electrical properties of composites based on conjugated polymers and conductive fillers. Carbon, 41(7), 1391-1395.
68. Tu, X., Xie, Q., Jiang, S., & Yao, S. (2007). Electrochemical quartz crystal impedance study on the overoxidation of polypyrrole–carbon nanotubes composite film for amperometric detection of dopamine. Biosensors and Bioelectronics, 22(12), 2819-2826.
69. Tyagi, M., & Tyagi, D. (2014). Polymer Nanocomposites and their Applications in Electronics Industry. Int. J. Electron. Electr. Eng, 7(6), 603-608.
70. Van Noorden, R. (2006). Moving towards a graphene world.
71. Xu, Y., Wang, Y., Liang, J., Huang, Y., Ma, Y., Wan, X., & Chen, Y. (2009). A hybrid material of graphene and poly (3, 4-ethyldioxythiophene) with high conductivity, flexibility, and transparency. Nano Research, 2(4), 343-348.
72. Wang, J., Musameh, M., & Lin, Y. (2003). Solubilization of carbon nanotubes by Nafion toward the preparation of amperometric biosensors. Journal of the American Chemical Society, 125(9), 2408-2409.
73. Wang, J. (2005). Carbon-nanotube based electrochemical biosensors: A review. Electroanalysis, 17(1), 7-14.

74. Zhang, Q., Rastogi, S., Chen, D., Lippits, D., & Lemstra, P. J. (2006). Low percolation threshold in single-walled carbon nanotube/high-density polyethylene composites prepared by melt processing technique. Carbon, 44(4), 778-785.

75. Zhu, N., Chang, Z., He, P., & Fang, Y. (2006). Electrochemically fabricated polyaniline nanowire modified electrode for voltammetric detection of DNA hybridization. Electrochimica Acta, 51(18), 3758-3762.

# CHAPTER 10

# APPLICATIONS OF COMPOSITE MATERIALS IN SPORTS INDUSTRY

## INTRODUCTION

Sports play an important role in the development and integration of social and cultural behaviors. Similar to the technological developments in science and technology, advancements in sports are also changing the lives of people. The developments and innovations in the modern sports industry require sports experts to pay special attention to research and development of the sports goods. Composite materials are an excellent choice for manufacturing of sports equipment because of their high strength to weight ratio along with other useful characteristics such as break resistance, fatigue resistance, superior thermal stability, abrasion resistance, friction resistance, vibration attenuation, flexible design and easy processing (Sutherland, 2004). Composite materials are being used extensively to manufacture multifold sports goods including the aquatic sports equipment, such as surfboards, sailboards, kayaks, sailing boats, gigs and planning boats; the ball game equipment, such as badminton rackets, tennis rackets, basketball stands backboards and softball bats; the winter sports goods, such as ice hockey sticks, ski poles snowboards and sledges; the poles utilized in high jump equipment, and the arrows and bows employed in archery equipment are all produced from composites instead of conventional materials such as wood, bamboo, plastics, and metals. Experiments have proved that composite sporting goods exhibit improved operational properties which help the athletes to achieve excellent performances (Froes, 1997). Table 15 shows the applications of different composite categories in sports goods.

**Table 15:** Applications of various composite structures in sports goods

| Form | Application |
|---|---|
| Platy Structure | Snowboard, surfboard, wind-surfing board, table tennis bat, spar of glider wing, etc. |
| Tubular Structure | Tennis racket, badminton racket, fishing rod, golf stick, softball bat, hockey stick, pole in high jump, arrow shaft, etc. |
| Flake Structure | Various helmets, golf stick head, body structure of various light boats |
| Other Structure | Various vehicles for race, swords, climbing ropes, various threads, etc. |

The boost in the economy is a source of improvement in the living standard of people; as the economy of a country flourishes, more and more people engage in different kinds of relaxing as well as recreational sports. With time, where, great emphasis has been paid to improve and develop the sports equipment, sports experts have also laid great emphasis on the training of athletes. Due to notable qualities of composite materials like high-strength, light weight, easy processing and large degrees of freedom of design, they have now widespread use in the manufacturing of sports equipment.

## RECENT TRENDS

Over the recent years, developed countries, by utilizing the benefits of material science and technology, has made remarkable progress in the application of composites in the sports industry. Different composite fibers like carbon fibers (CF), glass fibers (GF), ceramic fibers or aramid fibers have been used in the manufacturing of tennis rackets; laminated composite materials have been used to manufacture Ski sticks, etc. Similar is the case with golf, paddle, and hockey stick (Rosen & Caiazzo, 2002).

FRP sports equipment has made admirable progress in China. High modulus glass fibers have been made from FRP poles by Shanghai Institute. Similarly, Zhejiang Fuchunjiang has developed advanced composites to help in the manufacturing of water sports equipment for sports like rowing. These sports equipment were also exported worldwide and were used in international competitions. In a different range of composite materials, the most popular composite in the sports industry is fiber reinforced composite. It is the main composite used for the manufacturing of sports goods in the 21st century.

The optimal design of sports gear necessitates the application of various disciplines. These various disciplines will not only enhance the performance but also help in making user-friendly equipment from the viewpoint of avoiding any injury. Therefore, while designing sports equipment, various characteristics of materials needs to be taken into consideration. Some of these characteristics include density, toughness, strength, fatigue resistance, damping, and cost of material. To design the sports gear according to our requirements, often a mixture of different types of material is chose. These mixtures may consist of metals, polymers, ceramics, and composites. Following is the list of sports goods, where composites can have a high potential of usage:

- Vaulting Pole
- Archery equipment
- Canoes and Kayaks
- Javelin
- Bicycle frames (Chen et al., 1993)
- Golf & Polo sticks
- Hand glider frame
- Protective Sportswear
- Windsurfing boards

## BENEFITS OF FIBER REINFORCED COMPOSITES (FRPS) FOR SPORTS GOODS

As it is common knowledge that before the use of fiber reinforced composites in the manufacturing of sports gear, materials like steel, stainless steel, wood, aluminum alloy, etc. were used for this purpose (Zhang, 2015). Fiber reinforced composite material has numerous advantages over these materials in the following aspects:

## Light Weight

As most of the times, humans have to carry sports equipment (for example racket of tennis, bikes golf club rackets, Skis, etc.); therefore, it is better to manufacture sports gear lighter in weight if possible. Benefits of using fiber reinforced composite materials for this purpose are innumerable. For instance, when carbon fiber reinforced composites are used to manufacture golf club rackets, the mechanical properties of the metallic rod are improved by 30 % to 50%. Moreover, the weight is lighter as compared to the metal rod.

## Improved Mechanical performance

High-performance sports goods require excellent mechanical strength, high fatigue resistance, and enhanced impact resistance. FRP composites have outstanding mechanical properties suitable for manufacturing sports equipment. Carbon fiber reinforced composites have excellent strength, stiffness and impact resistance which makes them an ideal choice for high production of high-performance sporting goods. Additionally, composite materials have excellent shock absorption characteristics which is a primary requirement in most of the sports goods.

## Flexible Design and Robust Processing

Besides lightweight and extraordinary mechanical properties, composite materials attract the commercial sports industry due to their easy processing and flexible designs. Customized sporting structures can be manufactured from a range of fiber/matrix combinations to achieve desired weight, mechanical strength and geometric tolerance (Mufti, 2003). Composite materials are being manufactured by various processing techniques including Automated Fiber Placement (AFP), Pultrusion, Resin Transfer Molding (RTM) and Filament Winding techniques. Moreover, composite materials require the least maintenance due to their excellent mechanical properties and superior resistance to chemical attacks.

## Economic and Environmental Benefits

Environmental and economic factors play a vital role in the development of composite materials for sports goods. Composite materials are less susceptible to corrosion and environmental attacks which make them an ideal choice for long term applications (Abilash & Sivapragash, 2013). Moreover, composite materials don't produce any degradation products, unlike metals and plastics which produce corrosion products and carcinogenic vapors respectively. Consequently, composite materials have extended lifetimes, low maintenance costs and high environmental protection.

# APPLICATIONS

There are many applications of composite materials in the sports sector, some common examples are discussed below:

## World's Lightest Bike

Bugatti and Dutch bike manufacturer PG have revealed the world's lightest (and perhaps the fastest) bike. The bike is inspired by sports car of Bugatti's Chiron. The weight of the bike is just 11 pounds (Fig.92); credit goes to carbon fiber composites, which have been used in its manufacturing. As reported by *Architectural Digest*, almost 95 percent of the complete bike has been made by carbon fiber composites.

Most of the bikes which have large quantities of composites in their manufacturing use carbon fiber composite in their frame (Nelson, 1997). However, the Bugatti/PG design employs carbon fiber in the bike's rims, handlebar, fork, seat, seat post, brake, and crank. A German company made the wheels of the bike, whereas, its frame was built by Merelli. The fork of the PG Bugatti bike was designed by THM Carbones. Merelli has an experience of around 30 years with composite materials, and with this vast experience, it has designed, developed and built many outstanding high-performance carbon goods for Formula 1 and cycling industry.

**Figure 92:** World's lightest bike manufactured from carbon fiber composites

As stated by PG, all of the high-tensile strength parts of carbon fiber are made by passing through a method known as prepreg performance process, abbreviated as P3. Manufacturing of all parts has been completed in Germany. Prepreg materials and methods are used in handcrafting of all carbon components. Previously these materials were only used in the manufacturing of aeronautical industries and motorsport to assure the highest rigidity and the lightest weight possible of the product.

Moreover, according to PG, all cross sections of the frames of the bike have been made oversized to guarantee maximum rigidity. However, it is to be noted that a total of only 667 bikes will be manufactured and these bikes are not envisioned to be used on public roads.

## Composite Ski or Float Ski

Chisnell & Goode (1998) carried out research on the development of composite skis for enhanced speed and lightweight. Later on, the inventor of polish, Kris Rogus, has produced the famous "Float Ski." Float Ski is the first device of its kind which empowers its users to walk effortlessly on the surface of the water (Fig.93). As claimed by Rogus, the Float Ski is fully ready for commercial use.

Rogus, throughout his life, was a fan of windsurfing and rowing. Over the years, he kept trying to design a product that will allow him to walk on water. His first try was when he attached two kayak boats in his teenage. In 2003, he moved to London and spent several years there trying to convert his passion into a business.

The weight of each Float Ski is around 10 kg. All the mechanical parts of the Ski are hand-made and were developed and assembled in his workshop in London. Rogus worked with Tris Cokes, a surfing legend of UK, to develop the design of Ski by employing polystyrene (PS) foam and epoxy resin. However, all of the mechanical parts were manufactured using carbon fiber.

**Figure 93:** Lightweight composite ski manufactured from CFRPs (carbon fiber, epoxy and polystyrene)

According to Rogus, he experimented with different materials, resins, foams, and tools to make something which is both light in weight and fast enough for the water. He further claims that a skilled user can use Float Ski to stroll on water at about 5 – 6 km/ hour rate. Further, the addition of poles can allow the user to reach up to 7 -8 km/ hour.

In his interview to Evening Standard, Rogus has mentioned that Float Ski is the first device of its kind which is fully functional and has been designed to let its user stay on the water surface for up to two hours. He further suggested gently that people can rent Ski for half an hour, try it and take a picture along with it. Rogus asked for the permission of trying Ski from fifteen central parks of London, out of which, only Hyde Park supported his small, unconventional business during his tests. The ultimate target market of Rogus for his creation is the 3.2 million people in the United Kingdom who regularly involve themselves in water sports such as kayaking, canoeing, yachting, wind surfing, and power boating.

## World's Fastest Catamaran Constructed for America's Cup

Oracle team, USA, has disclosed the fastest catamaran (Fig.94) of the world for America's Cup. Jimmy Spithill, who is the leader of Oracle team USA, will lead his team out on the waters of Bermuda to preserve the America's Cup. The team has disclosed the catamaran just now that will be used in the competition. The catamaran is an AC50 model and is joined with the help of a lattice of carbon fiber tubing. Both Oracle team and Red Bull are claiming it as the world's fastest sailboat.

Since long, Oracle team has been known to use composite materials for the manufacturing of catamaran and other applications. In the winter of 2016, a crew was led by Spithill from Bermuda to New York in a total of mere 66 hours on a composite made catamaran famous with the name F4. According to Spithill, this journey helped him in improving his sailing skills and prepared him well for the America's Cup.

**Figure 94:** Composite based world's fastest Catamaran manufactured for America's Cup

The team has amalgamated with New Zealand founded Core Builders Composites for more than six years. Oracle Team USA have utilized their facility for their America's Cup projects which dates back to the 2010 America's Cup, and again in 2013 for the successful defense of the America's Cup.

In 2014, Oracle Team USA made an announcement of joining hands with Airbus on instrumentation, aerodynamics, composites, simulation, hydraulics, structures, and data analysis proficiency for the AC50. For optimization of performance, Core Builders Composites capitalized in creating methods and materials to reduce the cost for manufacturing composite tools. Tim Smyth, who is a manager at Core Builders Composites, has made it clear that they are using cheap mold manufacturing methods and inexpensive carbon fiber, which includes a large quantity of recycled carbon fiber as well to build all carbon molds. Carbon molds are preferable because they are durable (Pervaiz & Sain, 2003).

The main competitors of Oracle Team USA will be Land Rover Ben Ainslie Racing of the United Kingdom and Emirates Team, New Zealand. Both of these teams have announced manufacturing of their own AC50 catamarans to compete for the America's Cup. Core Builders Composites, in union with Cookson Boats, have constructed tools, and various team building parts from those tools for the AC50.

## Composite Smart Bikes Powered by Android

Extensive research is being carried out to manufacture smart bikes embedded with digital android systems (Kiefer & Behrendt, 2016). One of the most attention-grabbing innovations on exhibition at CES® 2017 in Las Vegas was two new "smart bikes" for the U.S. market, manufactured by LeEco. According to LeEco, these bikes are going to break the cycle of orthodox riding (Fig.95).

Each bike features LeEco's Andriod 6.0 powered bike operating system which is displayed on a 4-inch touch screen. The operating system is powered with the help of a 6000mAh rechargeable lithium-ion battery. The system provides a variety of features such as online and offline playback music, ride logging, turn-by-turn navigation personalized for bicycle riding, and walkie-talkie communication with any other compatible LeEco smart bikes present within a closed location.

**Figure 95:** Composite based digital smart bike manufactured by LeEco

The LeEco Smart Road Bike has also earned an award at the conference. This Smart Road Bike has an aerodynamic Toray T700 carbon fiber frame, seat post, fork, wheels and a handlebar. Some of the other components are a lightweight brake set which features dual-pivot technology and is paired with brake/shift levers, attached to lightweight and a sturdy handlebar made of carbon fiber. The weight of the bike is just 18.5 pounds due to extensive use of carbon fiber in its structure.

Another bike of LeEco, known as Smart Mountain Bike has a Toray T700 carbon fiber frame, a wheelset of 27.5″/650b diameter and a handlebar. The weight of this bike is 26.9 pounds. Although LeEco has not revealed the final price of these two models, both these smart bikes are expected to be available in the U.S. market by the end of 2017.

## Exclusive HP4 RACE Motorcycle Manufactured from Carbon Fibers

Carbon fiber is an excellent choice to manufacture lightweight and high-performance motorbikes (Bellanca, 1991). In Milan's Motorcycle Show (2017), a German company, BMW Motorrad revealed an advanced prototype of its BMW HP4 RACE. The company has announced this new motorcycle as its most exclusive motorcycle ever. Just like any other BMW vehicle, this motorcycle has a significant amount of carbon fiber in its structure. The 2017 HP4 RACE is the latest and the most advanced version of the HP4 which made its debut in 2013. HP4 was the first bike in the market to feature semi-active suspension system.

As announced by the president of BMW Motorrad, HP4 RACE will use composites not only in the manufacturing of its mainframe but also in its fiber rims. This motorcycle will join Ducati's Superleggera as the only bike in the market featuring very lightweight carbon frames. More information about this model is likely to be revealed in next spring.

It was stated on sportrider.com, that BMW had manufactured the main carbon fiber frame, wheels, self-sustaining tail section, and the whole bodywork in any of its three manufacturing sites located in Leipzig, Landshut and Dingofing, Germany. The site further claims that unlike other automakers in the market, who generally outsource the manufacturing components, BMW is one of the first auto manufacturing company which has made a substantial investment into its carbon fiber manufacturing abilities. An exclusive limited series of HP4 RACE will be manufactured and made available in the second half of 2017. The CFRP-based composite bike is shown in Fig.96.

**Figure 96:** High-performance composite motorbike manufactured by BMW Motorrad

## Leafboard's Electric Skateboard Constructed from Composites

Leafboard is a startup company which has manufactured a lightweight electric skateboard, affordable in price that lets its users skate effortlessly. Electric skateboards of the past usually had difficulty in handling inclines (Armstrong, 1989). The company has raised approximately $239,604 on its Kick starter page, which is more than its original goal of raising $50,000.

The weight of this electric skateboard is just 9.7 pounds (Fig.97). Leafboard claims that it is the lightest freestyle electric skateboard offered in today's market. The light weight of skateboard has been possible by employing three layers of fiberglass together with eight layers of Canadian maple, in the manufacturing of its deck, while 2.5 mm of carbon fiber has been used to manufacture the battery case of the board. According to the Kickstarter page of the company, once the company reaches its goal of $350,000 in funding, they will upgrade the board to feature a full carbon fiber composite deck.

**Figure 97:** Electric skateboard manufactured from glass fiber composites

This skateboard is powered by a lithium-polymer battery, having a capacity of 5000 mAh, and is rechargeable for more than 1000 times. The top speed at which the board reaches is 19 mph, and it has a range of almost 11.5 miles. The speed and braking of this skateboard are controllable with the help of a remote control. Moreover, it also has a USB port to charge a phone if required. For complete charging of iPhone 6, this board takes only 70 minutes, and charging will reduce the range of the board by mere of 1.25 miles. Leafboard has also stated that both the deck and motor of the skateboard have been processed with a distinctive treatment to make the board water resistant. A pre-order of Leafboard's skating board is just $369. However, its regular retail price will be $649.

Nevertheless, leafboard is not the only company which is making electric skateboards on commercial level. Back few years, another company, known as Marbel, launched an analogous product. Marbel's skateboard is 0.2 pounds heavier than Leafboard's skateboard and it has a slighter short range. The deck of Marbel's board has been manufactured using a combination of Kevlar and carbon fiber.

## 3D Printed Amphibious Fins

Presently, composites are used to develop various aquatic products, including amphibious fins via 3D printing techniques (Liljenhjerte & Kumar, 2015). A New York-based Composite Prototyping Center (CPC) Plainview, in partnership with Northwell Health, has created the first 3D-printed, prosthetic, amphibious leg. This leg is known as "The Fin." The leg enabled 33-year-old Dan Lasko who is a Marine Corps veteran amputee to swim again. According to Thomas Thornton, who is the senior vice president of Northwell Ventures, the prosthetic market is mainly characterized by its one size solutions which fit all. Previously there was no device in the market that was fully amphibious to help amputees swim who have a passion for swimming. Therefore Northwell Ventures created something that didn't exist and in doing so solved a particular problem in a very remarkable manner.

3-D print of the prosthetic was made on a Mark Forged printer. For printing purpose carbon fiber enhanced nylon was used; a special material that permits the prosthetic to have both the flexibility and the strength required for its performance. With the help of 3-D printing, it is easy to quickly design and manufacture any prosthetic like the swim fin, which makes it possible to tailor add-ons at a rational price. Hence, it can be said that using composites for the manufacturing of detachable leg has offered several advantages among which main is corrosion resistance.

The 3-D printed amphibious fin (Fig.98) created by Northwell Ventures is completely patient-specific. It provides the patient enough support and propulsion. Due to the large quantity of plastics and carbon fiber used in it, corrosion is not a problem. It has been tested many times sending in and out of the Atlantic Ocean and never any trace of corrosion has been found on it. According to an estimation, approximately 1.9 million people nationwide have lost a limb. As reported by the Amputee Coalition of the United States, this number is expected to get double by 2050; the main reason being diabetes. Northwell Ventures believes that prosthetics like amphibious fins can play an active role in helping amputees to recommence their active lifestyle again.

**Figure 98:** 3-D printed amphibious fin produced by Northwell Ventures

## Development of The World's Primary Autonomous Electric Racecar

During Roborace (Barcelona's Mobile World Congress), a startup company dedicated to creating autonomous vehicles revealed its Robocar. This Robocar is getting popular as the world's first self-driving electric race car. This car contains five light ranging and detecting radar sensors which are known as LiDAR (four of sensors are on the front, and one is on the back), 18 ultrasonic sensors, two radars, six artificial intelligence cameras, two optical speed sensors, and satellite navigation.

With this new racing car in the market, Roborace has opened a new dimension of sports where racing cars meet the unrelenting rise of artificial intelligence (Anderson et al., 2014). Four motors each having 300 kilowatts of power, and weighing 154 pounds have been used to power the Robocar. Further, a 540-kilowatt battery is used for its all-wheel drive during the race. The frame of the car has been primarily composed of carbon fiber composites. The weight of the Robocar is just 2150 pounds, and the maximum reachable speed of the car is 200 mph. Roborace gained expertise in composites for building cars from UK-based Lentus Composites. Lentus Composites have experience in the engineering of Formula 1 racing applications.

The CEO of Roborace, Denis Sverdlov, is also the chief executive officer of another company named as Charge. In November, Charge revealed a self-driving delivery van which is also made of composites. The goal of the company is to demonstrate that self-driving technology is both cool as well as feasible.

According to Sverdlov, Roborace is going to help people to accept robots on the road. The goal of the Roborace is to embrace this concept entirely to demonstrate that it will work flawlessly. Two Robocars (Fig.99) are expected to be fully completed by the end of this year and be used in Roborace's self-driving competitions.

**Figure 99:** Robocar developed by Barcelona's Mobile World Congress

## Composite Car Sets Guinness World Record for Fuel Efficiency

Few students from the Technical University of Munich, Germany have made their entrance into the Guinness World Records due to their vehicle which is known as eLi14. This electric and solar powered car has reached 26,135 miles per gallon. TU fast cars have been used in the Shell Eco Marathon since 2011. The weight of this vehicle is less than 70 pounds, and it has a wedge-like shape. The monocoque of the car is composed of carbon fiber reinforced plastic (CFRP). According to the team, the chassis of the vehicle was specifically made with CFRP prepreg materials, also containing ROHACELL® sandwich materials from specialist Evonik. The team has further informed that CERP tooling has been used with a Vacuum Assisted Process (VAP) to manufacture the chassis.

According to Philipp Wurdak, technical project manager of the team TUfast, "we have to take care when a car like three areas." It is important for a car to be streamlined to the maximum possible limit, maintain low rolling resistance, and build an optimally designed engine. Composite materials were also used in the manufacturing of some of the vehicle's running gear for example in custom-design and custom-made tubeless CFRP hollow chamber rims (Kresge & Lohse, 1996).

Lisa Kugler who is the driver of the model of the eLi14 stated that the vehicle was able to exceed the expectations of the team successfully. During the race in which the car broke the previous world records, the team ran into some problems when they found out that one of the technological components of the vehicle had been inserted wrongly. Nevertheless, the team quickly resolved the issue and proceeded to finish the race with record efficacy on the track at Audi. Electric and solar power composite car is shown in Fig.100.

**Figure 100:** Electric and solar powered car (eLi14) manufactured at TUM, Germany

## Safety Helmets for Baseball Players (Pitchers)

A star pitcher for the University of Hawaii, named Brendan Hornung, almost lost his life due to a skull fracture in 2014. The same man has now resumed his game and is now staying safe by using an extra resilient composite plate in his new helmet.

Hornung stated that since its carbon fiber it is quite strong. The company who manufactured the composite plate is Atlanta, Ga.-based Company, named Safer Sports Technologies (SST). The head guards of SST are made with CFRP, which is responsible for making them strong, stiff and lightweight and exceptionally safe. As stated by SST, these helmets can be trusted to significantly reduce any danger of severe brain injuries or skull fractures like the one suffered by Hornung in 2014.

While playing baseball, skull fractures are likely to occur at the exit velocity of the ball of almost 85 feet per second (equal to 58 mph). SST uses high collision impact tests to put its head guards. These tests give clear results demonstrating how composite technology can lower the force of impact.

Hornung is not the first pitcher who has benefited from composite protection. During Spring Training of Major League Baseball in 2016, almost around 20 pitchers tried using a hybrid of a helmet and a baseball cap known as the "Half Cap." The Half Cap was made by a company named Boombang, which is based in Los Angeles. It is styled in a belligerent manner and is made of carbon fiber composite shell together with a conforming foam liner and an innovative energy-absorbing impact layer. Another company based in Stamford USA has also started to manufacture composite baseball helmets.

## *Fastest Electric Vehicle*

An electric racing car has been built by students of an Academic Motorsports Club Zurich (AMZ) based in Switzerland. This racing car has exceeded the previous world record of fastest acceleration with a battery powered vehicle by covering distance at the rate of 0 – 62 miles per hour in just 1.513 seconds. The previous world record was set by a team from the University of Stuttgart, Germany at a rate of 0 – 62 mph in 1.779 seconds.

The weight of Grimsel (ETH Zurich's car) is merely 370 lbs. owing to the excessive use of carbon fiber in its manufacturing. A team comprising of 30 students from ETH Zurich and Lucerne University of Applied Sciences and Arts developed and built the car in less than a year. The main composite used is carbon fiber; additionally, the car has four-wheel drive with each of its wheel-hub motors having the capacity of generating 200 horsepower and weighs 1254 pounds. The performance of each wheel is regulated by a sophisticated traction control which allows the acceleration of the car to be increased even more.

Grimsel is the fifth electric vehicle of AMZ. It has set novel and new standards in lightweight manufacturing and electric drive technology. Previously there is not any large-scale production car (even cars with the combustion engine) which has been able to exceed the acceleration of Grimsel. Besides being the fastest car, Grimsel is also the most successful competition car. Formula Student is the most popular and the largest competition worldwide for engineers and is organized yearly on race courses around the world. Grimsel has abled AMZ to win the championship three times and has occupied the first position since 2013 in Formula Student world rankings.

**Figure 101:** High-speed sports car (Grimsel) constructed from carbon fiber composites at ETH Zurich

## Safer Football Helmets

For the last few years, the subject of helmet safety has been brought forward multiple times, with even the president deliberating the issue. Now a collaboration has been formed between bioengineering and packaging researchers at Clemson University and two companies based in North Carolina to deliver more advanced helmet materials to the sports industry. The researchers are actively exploring the efficiency of Innegra H yarns (hybridized yarns containing Innegra). The Innegra yarns join high strength fibers with Innegra S fibers to produce hybrid yarns which contain unique features. The Innegra H yarns have been found to excessively rising the impact resistance of fiber reinforced composites which has resulted in stronger and safer composite structures. This research is likely to change the currently used materials to manufacture football helmets.

The Clemson scientists are functioning with B&W Fiber Glass of Shelby, N.C., and Innegra Technologies of Greenville, N.C. to assess if Innegra-based composites can offer enhanced head protection to players who are involved in contact sports; for example, football. High impact applications and sports like kayaks and hockey already use Innegra-based composites. If proven effective, the Innegra H yarns can be expected to be employed in various applications where the top priority is the safety like automobiles and sports equipment.

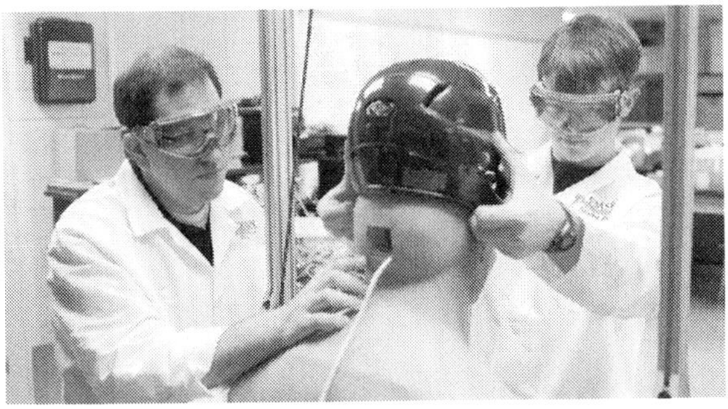

**Figure 102:** Football helmet manufactured from hybrid fibers (Carbon, Innegra S & Innegra H)

Previously, nearly all the developments in head protection have focused only on adding padding or making adjustments in the helmet's shape. Very few improvements have been made to the materials employed in the helmet's shell. The study mentioned above, is, therefore, capable of altering the entire thought process of designing the high-impact head protection helmets and the materials used in their manufacturing.

## Composite Snowboards for Downhill Sports

Snowboarders while riding down precarious slopes rely on some plastic strapping and a thin fiberglass piece. During tough competitions, snowboarders should be able to trust the equipment they are going to use to ride down mountain slopes to keep them from breaking. These boards are tested in extreme conditions during the ESPN Winter X-Games. With the manufacturing of advanced composites snowboards, the competition between snowboarders is likely to increase now more than ever.

K2 Sport located in Seattle has been designing snow skis composed of composite materials for the last 50 years. This company was established in November 1961. The founder of the company, Bill Kirshner, is a pioneer of fiberglass manufacturing in sports equipment. Before Kirshner, most of the Ski manufacturers used metal or wood to make components of the ski. Kishner who previously was a manufacturer of fiberglass cages, employed composites in the manufacturing of winter sports gear knowing that using fiberglass could prevent corrosion and other wood rot problems concomitant with traditional winter sports equipment. K2 Snowboards, an exclusive branch of K2 Sports dedicated to the manufacturing of composite snowboards, uses hand lay-up to manufacture high-performance sports gear. K2 Snowboards has made a significant investment in research and development of snowboard technology and is highly capable of delivering 100% production level snowboards in just four business days.

K2 Snowboards manufactures several thousand snowboards and more than 200 prototype boards every year. Mathematically, this is equivalent to manufacturing one board every day which they mold to ride, break and then make it better. The research facility at K2 Snowboards provides its manufacturer the opportunity to enhance the engineering design of the board before testing. Although some snowboards have started to implement closed molding technology to enhance the speed of the production, still, hand lay-up method is necessary for manufacturing the high-performance boards.

Figure 103: Manufacturing of composite snowboard at K2 Snowboards

# FUTURE TRENDS

For many years, composite materials have been enhancing the performance of various sports goods including Hockey sticks (adjustment in kick point to enhance puck velocity and energy transfer), fishing rods (reducing their weight and increasing their strength), Tennis rackets (reducing the weight of the racket and increasing its sweet spot size), Canoes (reduction in weight) and arrows (decreasing the energy loss by increasing their stiffness). It is expected that in future, the application of the fibrous material in sports gear will continue to evolve. The main research areas in the near future are:

- To enhance the performance and comfort level of athlete by employing new and advanced fibrous materials and structures
- To observe the vital signals
- To use fibrous sensors to evaluate the performance of the athlete
- Use of biomimetic to obtain solutions that offer certain improvements in performance (reducing friction, weight, aero, and hydrodynamics)
- To find recycling methods of composite materials to reduce associated environmental problems.

A design studio, named AL_A, and supervised by a famous architect Amanda Levete has revealed a new concept of tower design known as Pitch/Pitch. This project, in collaboration with Arup engineering company, aims to stack various soccer field on the top of each other. The design is created as a reaction to central London's lack of practicable inner city sports venues.

**Figure 104:** Design of a stackable soccer field constructed from composite materials

As stated by Levete and her team, the structure of the system would be modular and composed of carbon fiber composite. According to AL_A, these towers would be easy and quick to install and could be used for huge events like World Cup tournaments. The expected life span of these towers ranges from merely few days to a year which means that they can be employed to fill the empty sites awaiting development.

In addition to taking advantage of reduced weight and the relative ease of construction properties of carbon fiber, the designers of the towers chose carbon fiber due to its aesthetic appeal. As stated by AL_A director

Maximiliano Arrocet, this intervention has been designed to enhance the ambiance of the theater and to encourage the spectators and potential players. Pitch/Pitch is unique in its feature of adapting to the number of players and enables the game to preserve an urban flavor.

## REFERENCES


1. Abilash, N., & Sivapragash, M. (2013). Environmental benefits of eco-friendly natural fiber reinforced polymeric composite materials. Int J Appl Innov Eng Manag (IJAIEM), 2, 53-59.
2. Anderson, J. M., Nidhi, K., Stanley, K. D., Sorensen, P., Samaras, C., & Oluwatola, O. A. (2014). Autonomous vehicle technology: A guide for policymakers. Rand Corporation.
3. Armstrong, R. B. (1989). U.S. Patent No. 4,795,181. Washington, DC: U.S. Patent and Trademark Office.
4. Bellanca, A. T., & Bellanca Jr, A. T. (1991). U.S. Patent No. 5,011,172. Washington, DC: U.S. Patent and Trademark Office.
5. Chen, J. P., Hwang, J. L., Jang, R. G., Wu, S. L., & Yu, R. H. (1993). U.S. Patent No. 5,271,784. Washington, DC: U.S. Patent and Trademark Office.
6. Chisnell, W. R., & Goode, D. P. (1998). U.S. Patent No. 5,759,664. Washington, DC: U.S. Patent and Trademark Office.
7. Froes, F. H. (1997). Is the use of advanced materials in sports equipment unethical? JOM, 49(2), 15-19.
8. Kiefer, C., & Behrendt, F. (2016). Smart e-bike monitoring system: real-time open source and open hardware GPS assistance and sensor data for electrically-assisted bicycles. IET Intelligent Transport Systems, 10(2), 79-88.
9. Kresge, E. N., & Lohse, D. J. (1996). U.S. Patent No. 5,576,372. Washington, DC: U.S. Patent and Trademark Office.
10. Liljenhjerte, J., & Kumar, S. (2015). Pull-out performance of 3d printed composites with embedded fins on the fiber. In MRS Proceedings (Vol. 1800, pp. mrss15-2135597). Cambridge University Press.
11. Mufti, A. A. (2003). FRPs and FOSs lead to innovation in Canadian civil engineering structures. Construction and Building Materials,

17(6), 379-387.

12. Nelson, R. H., Milovich, D., Wilcox, W. M., & Read, R. F. (1997). U.S. Patent No. 5,624,519. Washington, DC: U.S. Patent and Trademark Office.
13. Pervaiz, M., & Sain, M. M. (2003). Carbon storage potential in natural fiber composites. Resources, conservation and Recycling, 39(4), 325-340.
14. Rosen, B. W., & Caiazzo, A. A. (2002). U.S. Patent No. 6,416,432. Washington, DC: U.S. Patent and Trademark Office.
15. Sutherland, T. W. (2004). U.S. Patent No. 6,755,757. Washington, DC: U.S. Patent and Trademark Office.
16. Zhang, L. (2015, April). The Application of Composite Fiber Materials in Sports Equipment. In 2015 International Conference on Education, Management, Information and Medicine. Atlantis Press.

# INDEX

## A



## B



## C



## D

## E



## F



## G



## I



## L



## M



## N



## O



## P